MW01174561

Lecture Notebook

for

Biochemistry

Fifth Edition

Mary K. Campbell
Mount Holyoke College

Shawn O. Farrell
Colorado State University

THOMSON

BROOKS/COLE

Australia • Canada • Mexico • Singapore • Spain • United Kingdom • United States

Cover image: © Digital Art/CORBIS

Printed in the United States of America
1 2 3 4 5 6 7   09 08 07 06 05

Printer: Thomson/West

0-534-40525-8

For more information about our products,
contact us at:
Thomson Learning Academic Resource Center
1-800-423-0563

For permission to use material from this text or
product, submit a request online at
http://www.thomsonrights.com.
Any additional questions about permissions can be
submitted by email to **thomsonrights@thomson.com.**

Thomson Higher Education
10 Davis Drive
Belmont, CA 94002-3098
USA

Asia (including India)
Thomson Learning
5 Shenton Way
#01-01 UIC Building
Singapore 068808

Australia/New Zealand
Thomson Learning Australia
102 Dodds Street
Southbank, Victoria 3006
Australia

Canada
Thomson Nelson
1120 Birchmount Road
Toronto, Ontario M1K 5G4
Canada

UK/Europe/Middle East/Africa
Thomson Learning
High Holborn House
50–51 Bedford Row
London WC1R 4LR
United Kingdom

Latin America
Thomson Learning
Seneca, 53
Colonia Polanco
11560 Mexico
D.F. Mexico

Spain (including Portugal)
Thomson Paraninfo
Calle Magallanes, 25
28015 Madrid, Spain

Lecture Notebook for
Biochemistry, Fifth Edition

Image Number	Figure	Description
1–3	1.11	A comparison of (a) an animal cell, (b) a plant cell, and (c) a prokaryotic cell
4	Table 1.1	Functional groups of biochemical importance
5	2.13 (bottom)	Titration of acetic acid with NaOH
6	2.13 (top)	Relative abundance of acetic acid and acetate ion during a titration
7–8	2.15	Titration curve and buffering action
9	Table 2.8	Acid and base forms of some buffers
10	3.1	Amino acid structure
11–12	3.6, 3.7	Titration curves of amino acids
13	Table 3.1	Common amino acids and their abbreviations
14	Table 3.2	PK_a values of common amino acids
15	4.4	β-pleated sheet
16	4.13	Forces that stabilize tertiary structure in proteins
17	4.23	Denaturation of proteins
18–19	5.5	Gel-filtration chromatography
20	5.6	Affinity chromatography
21	5.9	Ion-exchange chromatography using a cation exchanger
22	5.14	Strategy for determining the primary structure of a given protein
23	5.20	Sequencing of peptides by the Edman method
24	6.1	Activation energy profiles
25	6.10	Lineweaver-Burk double-reciprocal plot
26	6.12	Lineweaver-Burk double-reciprocal plot for competitive inhibition
27	6.13	Lineweaver-Burk double-reciprocal plot for noncompetitive inhibition
28	7.4	Monod-Wyman-Changeux model
29	7.5	Monod-Wyman-Changeux (or concerted) model
30	7.7	Sequential model of cooperative binding of substrate to allosteric enzyme
31	7.8	Phosphorylation of the sodium-potassium pump
32	7.9	Glycogen phosphorylation activity
33–34	7.14	The mechanism of chymotrypsin action
35	8.5	Structures of some phosphoacylglycerols
36	8.11	Lipid bilayer asymmetry
37	8.18	Fluid-mosaic model for membrane structure
38	8.24	The sodium-potassium ion pump
39–40	9.4	Commonly occurring nucleotides
41	9.5	A fragment of an RNA chain
42	9.6	A portion of a DNA chain
43	9.7	The double helix
44	9.19	Types of RNA

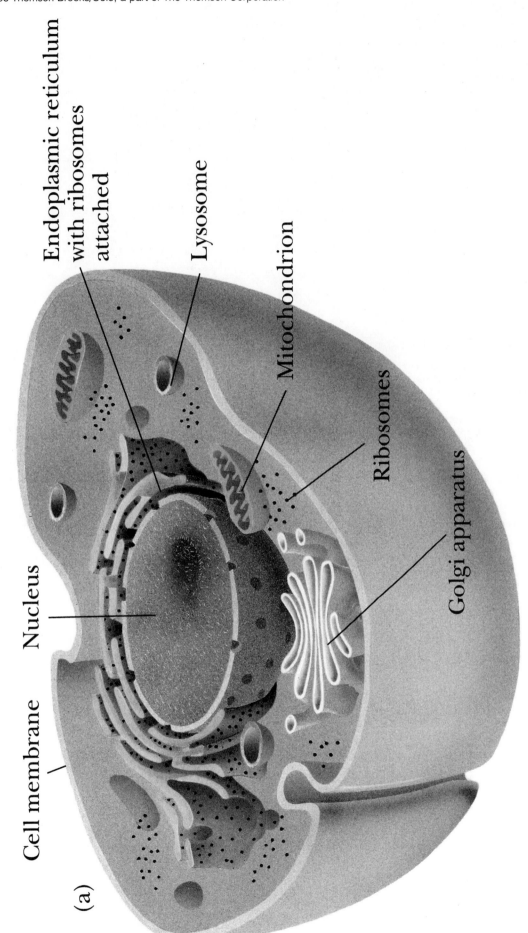

1

Endoplasmic reticulum with ribosomes attached

Lysosome

Mitochondrion

Ribosomes

Golgi apparatus

Nucleus

Cell membrane

(a)

Figure 1.11a A typical animal cell

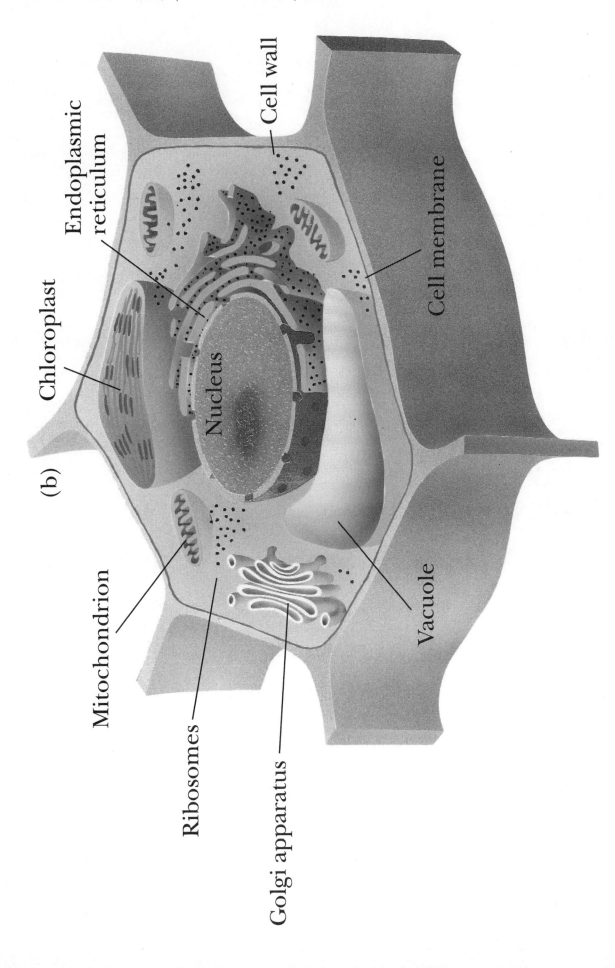

Figure 1.11b A typical plant cell

Cell wall

Endoplasmic reticulum

Cell membrane

Chloroplast

Nucleus

(b)

Mitochondrion

Ribosomes

Vacuole

Golgi apparatus

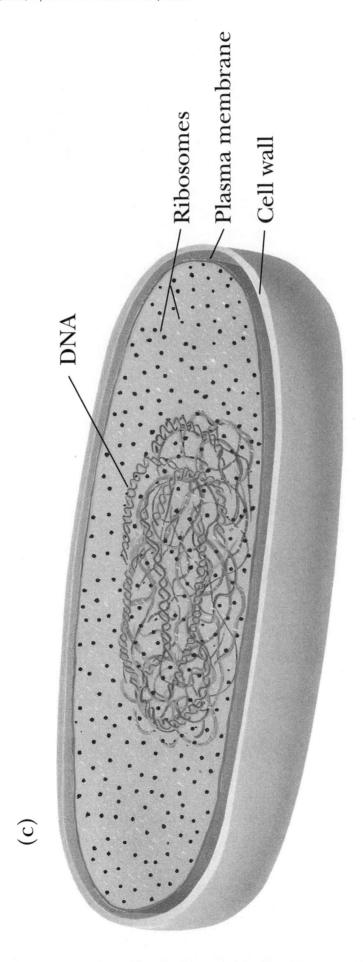

Ribosomes

Plasma membrane

Cell wall

DNA

(c)

Figure 1.11c A prokaryotic cell

Table 1.1

Functional Groups of Biochemical Importance

Class of Compound	General Structure	Characteristic Functional Group	Name of Functional Group	Example
Alkenes	$RCH{=}CH_2$ $RCH{=}CHR$ $R_2C{=}CHR$ $R_2C{=}CR_2$	$C{=}C$	Double bond	$CH_2{=}CH_2$
Alcohols	ROH	$-OH$	Hydroxyl group	CH_3CH_2OH
Ethers	ROR	$-O-$	Ether group	CH_3OCH_3
Amines	RNH_2 R_2NH R_3N	$-N{<}$	Amino group	CH_3NH_2
Thiols	RSH	$-SH$	Sulfhydryl group	CH_3SH
Aldehydes	$R-\overset{\overset{O}{\|\|}}{C}-H$	$-\overset{\overset{O}{\|\|}}{C}-$	Carbonyl group	$CH_3\overset{\overset{O}{\|\|}}{C}H$
Ketones	$R-\overset{\overset{O}{\|\|}}{C}-R$	$-\overset{\overset{O}{\|\|}}{C}-$	Carbonyl group	$CH_3\overset{\overset{O}{\|\|}}{C}\,CH_3$
Carboxylic acids	$R-\overset{\overset{O}{\|\|}}{C}-OH$	$-\overset{\overset{O}{\|\|}}{C}-OH$	Carboxyl group	$CH_3\overset{\overset{O}{\|\|}}{C}\,OH$
Esters	$R-\overset{\overset{O}{\|\|}}{C}-OR$	$-\overset{\overset{O}{\|\|}}{C}-OR$	Ester group	$CH_3\overset{\overset{O}{\|\|}}{C}\,OCH_3$
Amides	$R-\overset{\overset{O}{\|\|}}{C}-NR_2$ $R-\overset{\overset{O}{\|\|}}{C}-NHR$ $R-\overset{\overset{O}{\|\|}}{C}-NH_2$	$-\overset{\overset{O}{\|\|}}{C}-N{<}$	Amide group	$CH_3\overset{\overset{O}{\|\|}}{C}\,N(CH_3)_2$
Phosphoric acid esters	$R-O-\overset{\overset{O}{\|\|}}{\underset{\underset{OH}{\|}}{P}}-OH$	$-O-\overset{\overset{O}{\|\|}}{\underset{\underset{OH}{\|}}{P}}-OH$	Phosphoric ester group	$CH_3-O-\overset{\overset{O}{\|\|}}{\underset{\underset{OH}{\|}}{P}}-OH$
Phosphoric acid anhydrides	$R-O-\overset{\overset{O}{\|\|}}{\underset{\underset{OH}{\|}}{P}}-O-\overset{\overset{O}{\|\|}}{\underset{\underset{OH}{\|}}{P}}-OH$	$-\overset{\overset{O}{\|\|}}{\underset{\underset{OH}{\|}}{P}}-O-\overset{\overset{O}{\|\|}}{\underset{\underset{OH}{\|}}{P}}-$	Phosphoric anhydride group	$HO-\overset{\overset{O}{\|\|}}{\underset{\underset{OH}{\|}}{P}}-O-\overset{\overset{O}{\|\|}}{\underset{\underset{OH}{\|}}{P}}-OH$

The symbol R refers to any carbon-containing group. When there are several R groups in the same molecule, they may be different groups or they may be the same.

Table 1.1 Functional groups of biochemical importance

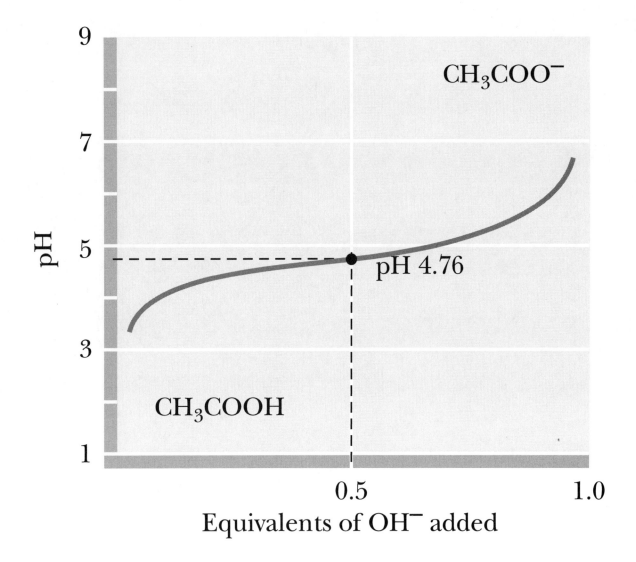

Figure 2.13 (bottom) Titration of acetic acid with NaOH

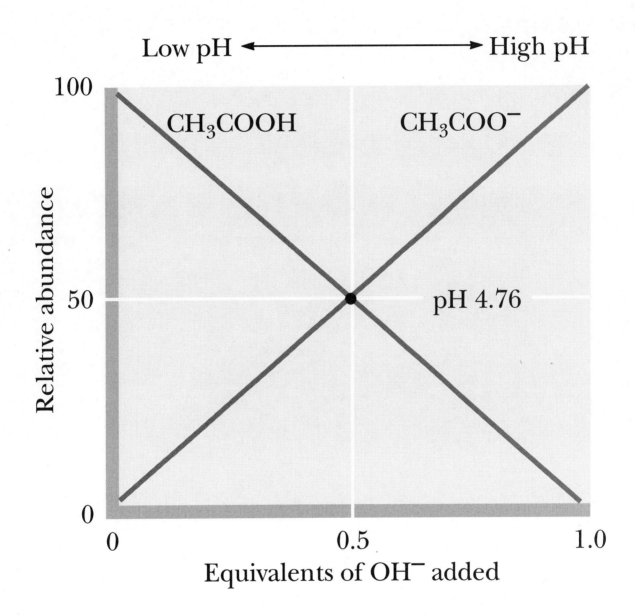

Figure 2.13 (top) Relative abundance of acetic acid and acetate ion during a titration

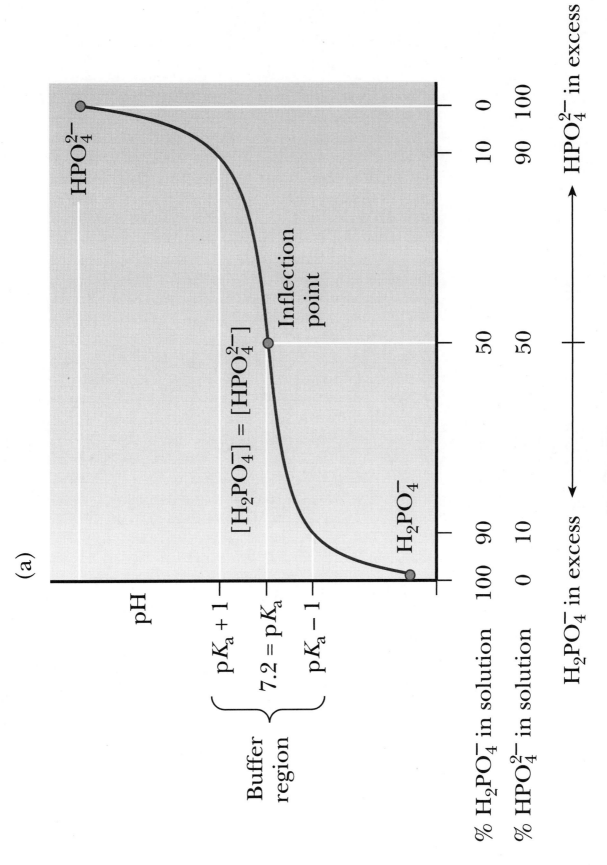

(a)

Figure 2.15a Titration curve and buffering action

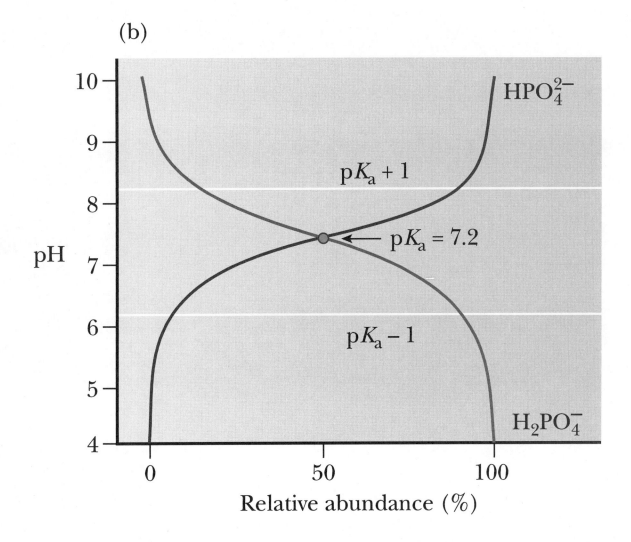

Figure 2.15b Titration curve and buffering action

Table 2.8

Acid and Base Forms of Some Useful Biochemical Buffers

Acid Form	Base Form	pK_a
N—$tris$[hydroxymethyl]aminomethane (TRIS) TRIS—H$^+$ (protonated form) $(HOCH_2)_3CNH_3^+$ ⇌	TRIS (free amine) $(HOCH_2)_3CNH_2$	8.3
N—$tris$[hydroxymethyl]methyl-2-aminoethane sulfonate (TES) $^-$TES—H$^+$ (zwitterionic form) $(HOCH_2)_3CNH_2^+CH_2CH_2SO_3^-$ ⇌	$^-$TES (anionic form) $(HOCH_2)_3CNHCH_2CH_2SO_3^-$	7.55
N—2—hydroxyethylpiperazine-N′-2-ethane sulfonate (HEPES) $^-$HEPES—H$^+$ (zwitterionic form) HOCH$_2$CH$_2$N$^+$—NCH$_2$CH$_2$SO$_3^-$ (with H) ⇌	$^-$HEPES (anionic form) HOCH$_2$CH$_2$N—NCH$_2$CH$_2$SO$_3^-$	7.55
3—[N—morpholino]propane-sulfonic acid (MOPS) $^-$MOPS—H$^+$ (zwitterionic form) O—$^+$NCH$_2$CH$_2$CH$_2$SO$_3^-$ (with H) ⇌	$^-$MOPS (anionic form) O—NCH$_2$CH$_2$CH$_2$SO$_3^-$	7.2
Piperazine—N,N′-bis[2-ethanesulfonic acid] (PIPES) $^{2-}$PIPES—H$^+$ (protonated dianion) $^-$O$_3$SCH$_2$CH$_2$N—NCH$_2$CH$_2$SO$_3^-$ (with H) ⇌	$^{2-}$PIPES (dianion) $^-$O$_3$SCH$_2$CH$_2$N—NCH$_2$CH$_2$SO$_3^-$	6.8

Table 2.8 Acid and base forms of some buffers

(a)

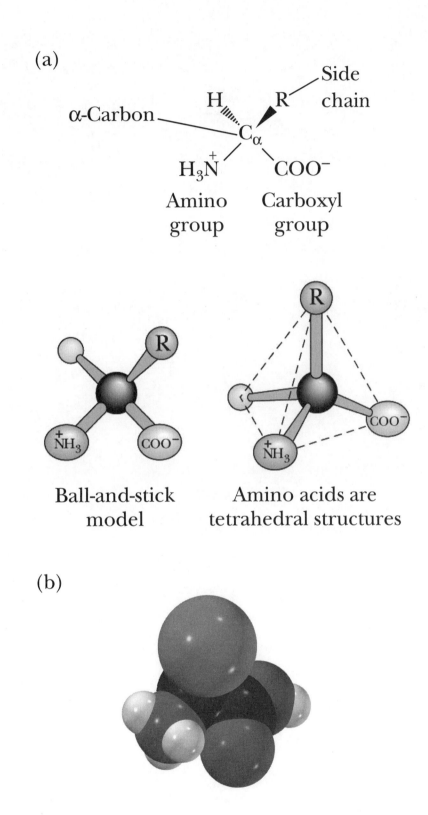

Ball-and-stick
model

Amino acids are
tetrahedral structures

(b)

Figure 3.1 Amino acid structure

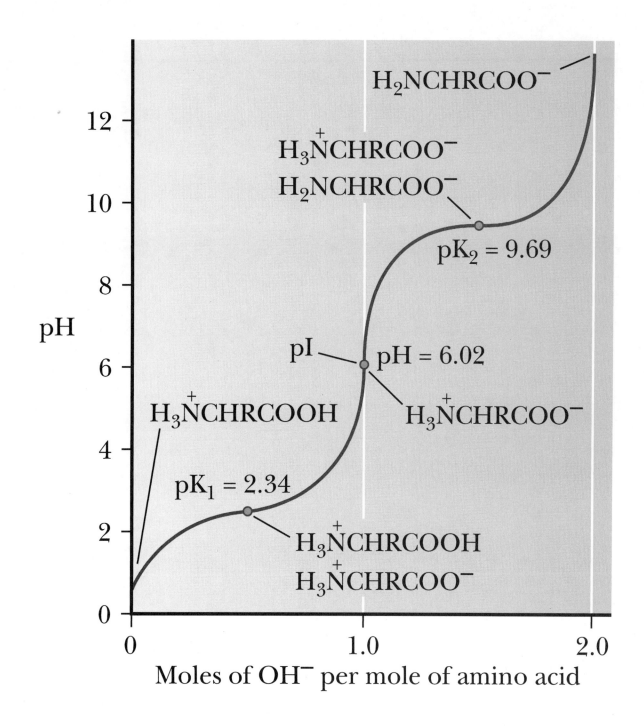

Figure 3.6 Titration curves of amino acids: alanine

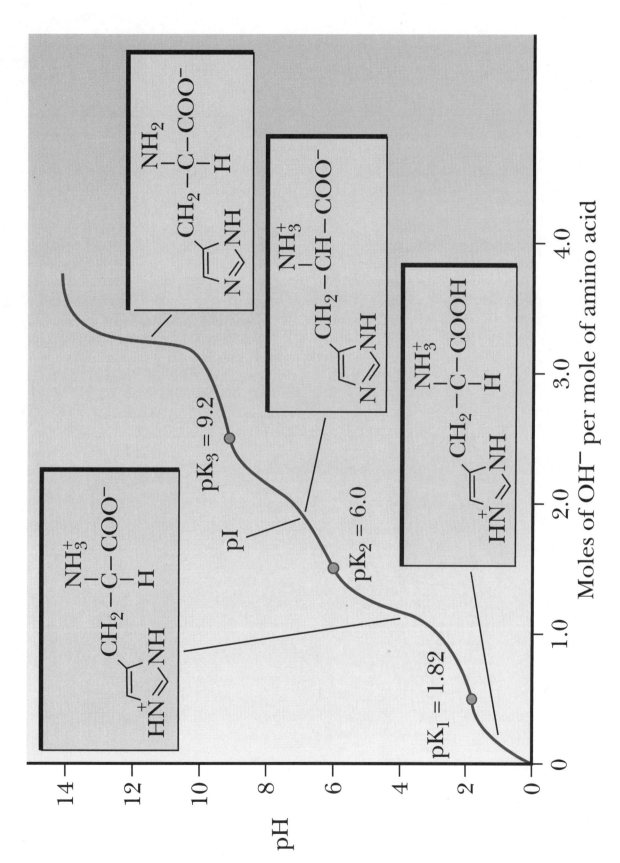

Figure 3.7 Titration curves of amino acids: histidine

Table 3.1

Names and Abbreviations of the Common Amino Acids

Amino Acid	Three-Letter Abbreviation	One-Letter Abbreviation
Alanine	Ala	A
Arginine	Arg	R
Asparagine	Asn	N
Aspartic acid	Asp	D
Cysteine	Cys	C
Glutamic acid	Glu	E
Glutamine	Gln	Q
Glycine	Gly	G
Histidine	His	H
Isoleucine	Ile	I
Leucine	Leu	L
Lysine	Lys	K
Methionine	Met	M
Phenylalanine	Phe	F
Proline	Pro	P
Serine	Ser	S
Threonine	Thr	T
Tryptophan	Trp	W
Tyrosine	Tyr	Y
Valine	Val	V

Note: One-letter abbreviations start with the same letter as the name of the amino acid where this is possible. When the names of several amino acids start with the same letter, phonetic names (occasionally facetious ones) are used, such as Rginine, asparDic, Fenylalanine, tWyptophan. Where two or more amino acids start with the same letter, it is the smallest one whose one-letter abbreviation matches its first letter.

Table 3.1 Common amino acids and their abbreviations

Table 3.2

pK_a Values of Common Amino Acids

Acid	α-COOH	α-NH$_3^+$	RH or RH$^+$
Gly	2.34	9.60	
Ala	2.34	9.69	
Val	2.32	9.62	
Leu	2.36	9.68	
Ile	2.36	9.68	
Ser	2.21	9.15	
Thr	2.63	10.43	
Met	2.28	9.21	
Phe	1.83	9.13	
Trp	2.38	9.39	
Asn	2.02	8.80	
Gln	2.17	9.13	
Pro	1.99	10.6	
Asp	2.09	9.82	3.86*
Glu	2.19	9.67	4.25*
His	1.82	9.17	6.0*
Cys	1.71	10.78	8.33*
Tyr	2.20	9.11	10.07
Lys	2.18	8.95	10.53
Arg	2.17	9.04	12.48

*For these amino acids, the R group ionization occurs before the α-NH$_3^+$ ionization.

Table 3.2 PK$_a$ values of common amino acids

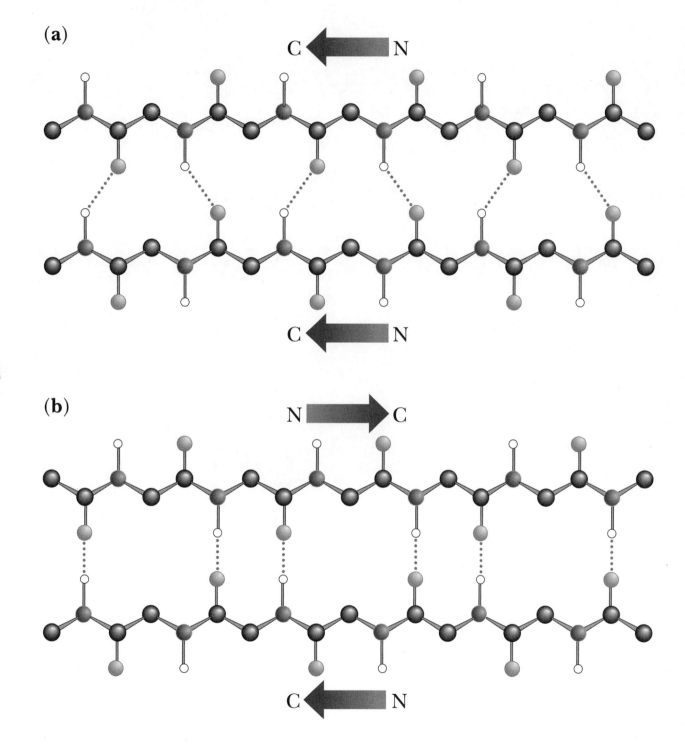

Figure 4.4 β-pleated sheet

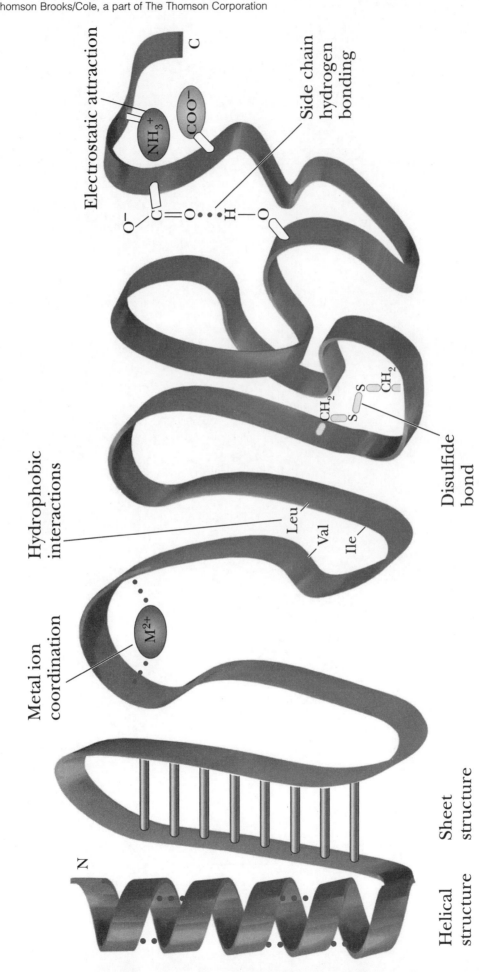

Figure 4.13 Forces that stabilize tertiary structure in proteins

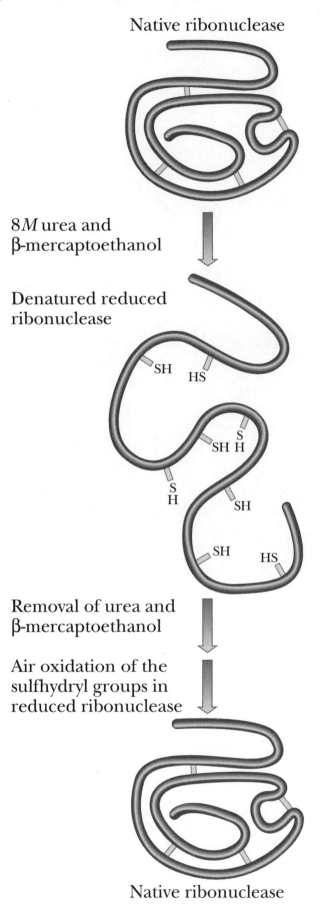

Native ribonuclease

8M urea and
β-mercaptoethanol

Denatured reduced
ribonuclease

Removal of urea and
β-mercaptoethanol

Air oxidation of the
sulfhydryl groups in
reduced ribonuclease

Native ribonuclease

Figure 4.23 Denaturation of proteins

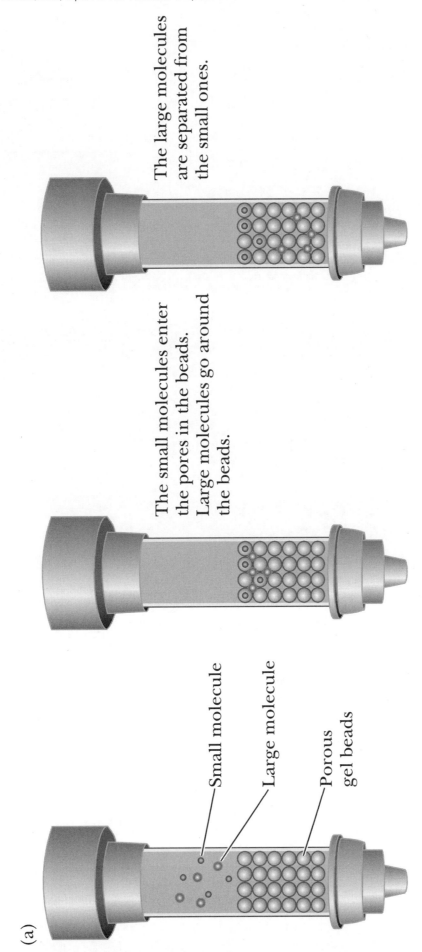

The large molecules are separated from the small ones.

The small molecules enter the pores in the beads. Large molecules go around the beads.

Small molecule

Large molecule

Porous gel beads

(a)

Figure 5.5a Gel-filtration chromatography

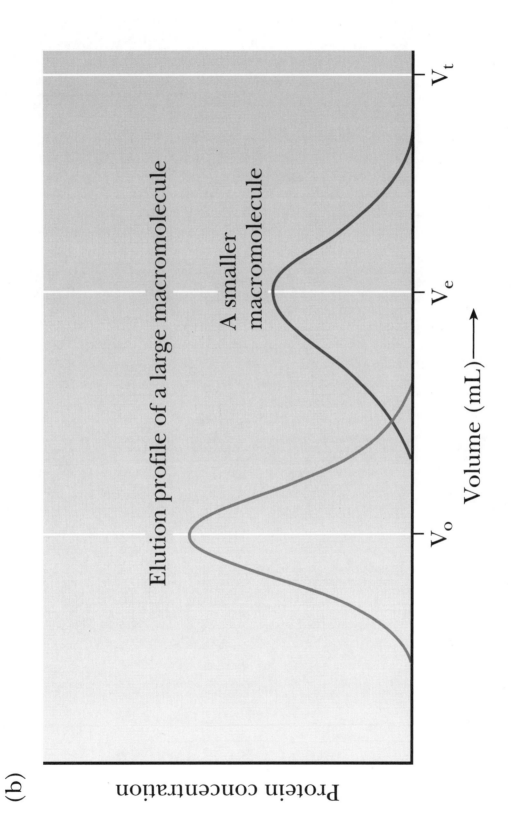

19

Figure 5.5b *Gel-filtration chromatography*

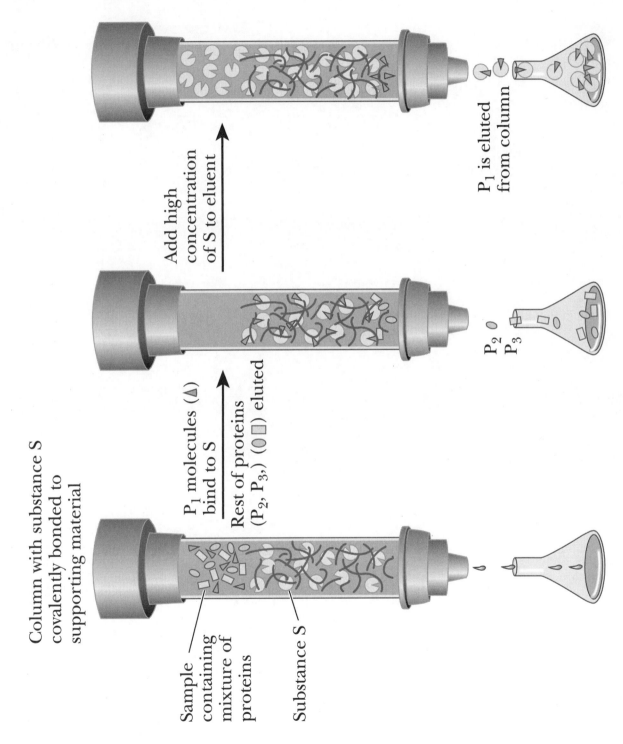

Column with substance S
covalently bonded to
supporting material

Sample
containing
mixture of
proteins

Substance S

P₁ molecules (△)
bind to S

Rest of proteins
(P₂, P₃,) (○□) eluted

Add high
concentration
of S to eluent

P₂
P₃

P₁ is eluted
from column

Figure 5.6 Affinity chromatography

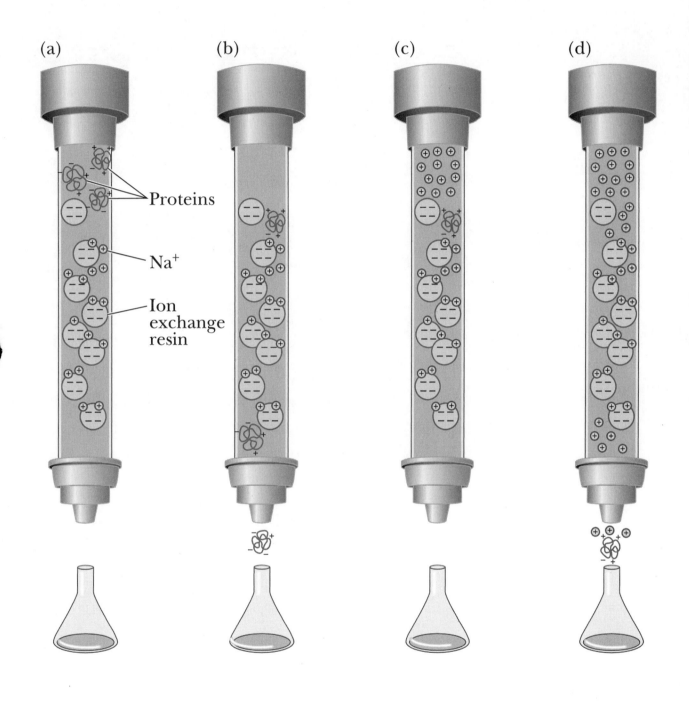

Figure 5.9 Ion-exchange chromatography using a cation exchanger

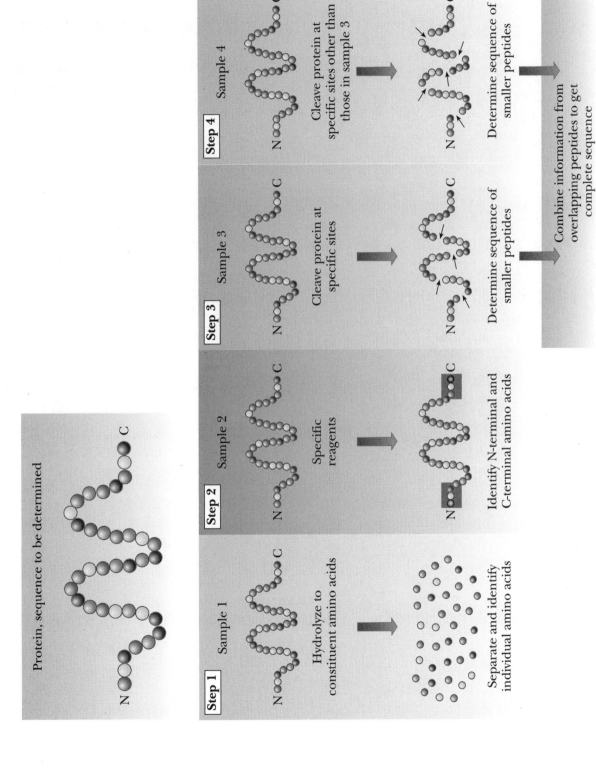

Figure 5.14 Strategy for determining the primary structure of a given protein

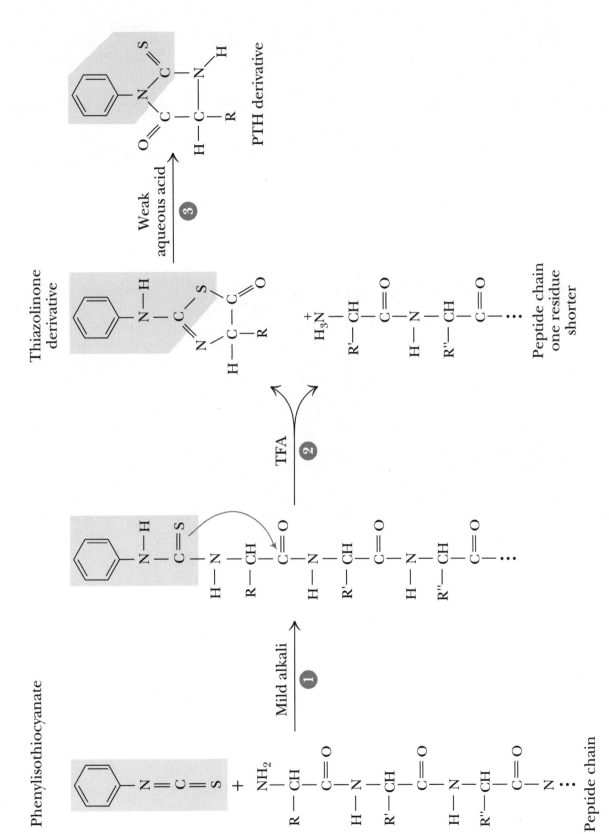

Figure 5.20 Sequencing of peptides by the Edman method

(a)

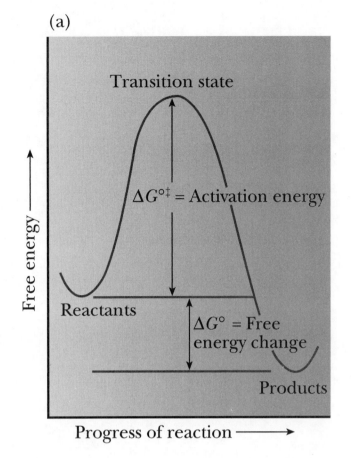

(b)

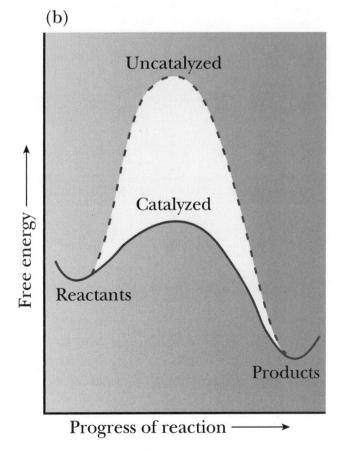

Figure 6.1 Activation energy profiles

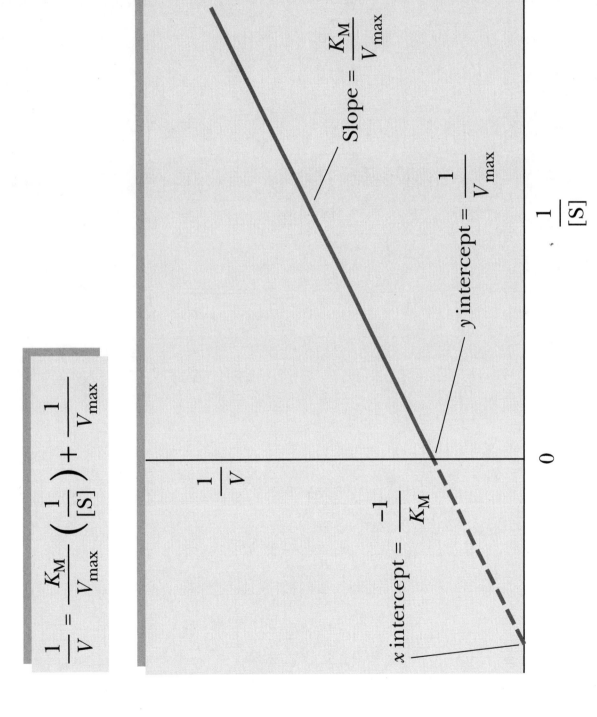

$$\frac{1}{V} = \frac{K_M}{V_{max}}\left(\frac{1}{[S]}\right) + \frac{1}{V_{max}}$$

Figure 6.10 Lineweaver-Burk double-reciprocal plot

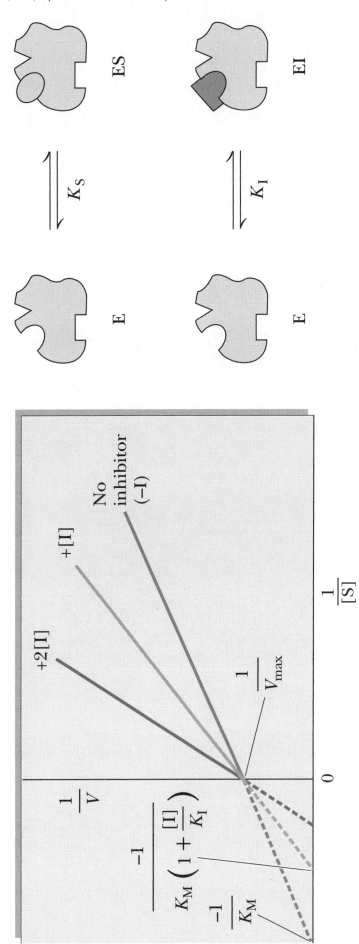

Figure 6.12 Lineweaver-Burk double-reciprocal plot for competitive inhibition

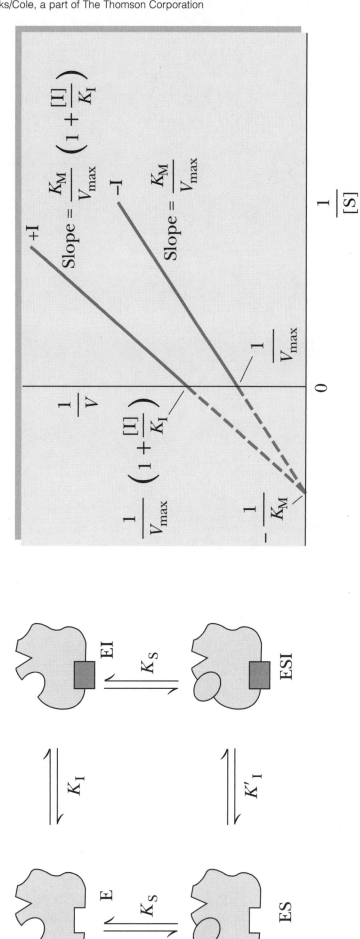

Figure 6.13 *Lineweaver-Burk double-reciprocal plot for noncompetitive inhibition*

(a) A dimeric protein can exist in either of two conformational states at equilibrium.

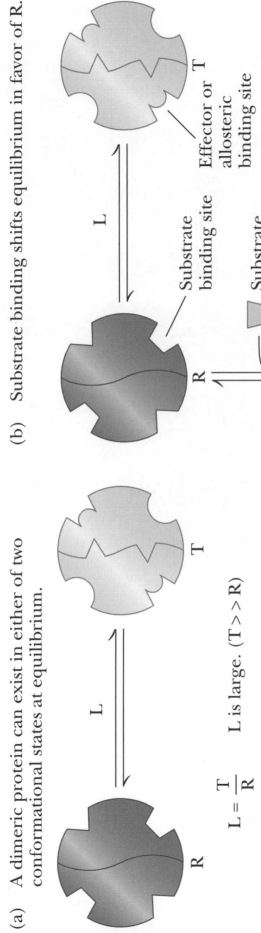

$$L = \frac{T}{R}$$ L is large. (T >> R)

(b) Substrate binding shifts equilibrium in favor of R.

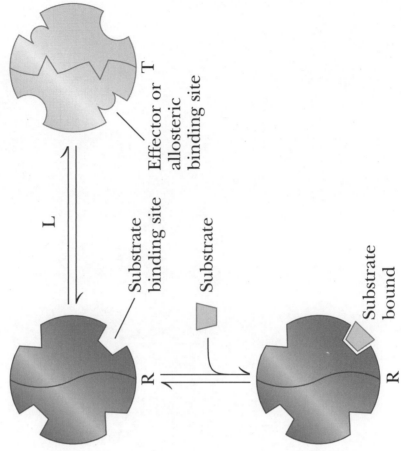

Effector or allosteric binding site

Substrate binding site

Substrate

Substrate bound

Figure 7.4 Monod-Wyman-Changeux model

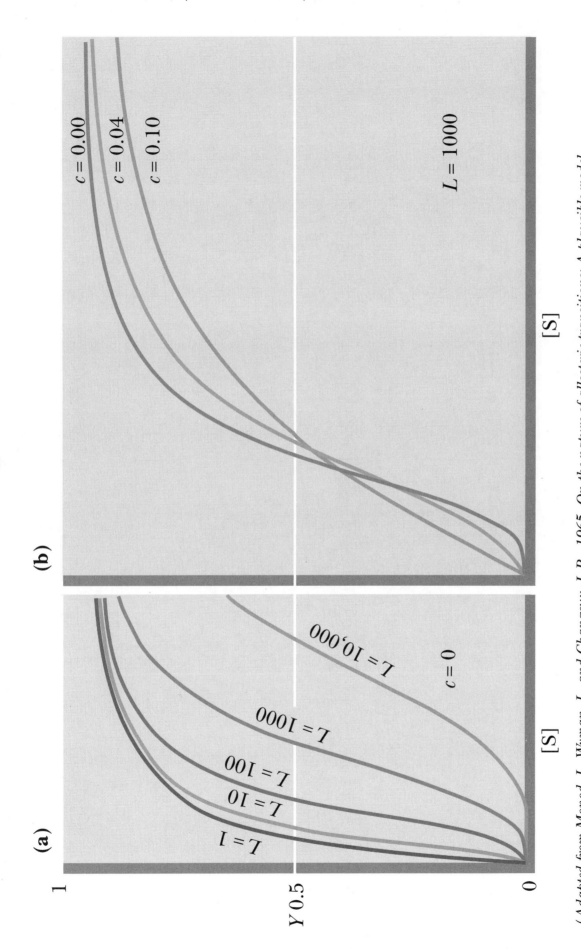

Figure 7.5 Monod-Wyman-Changeux (or concerted) model

(*Adapted from Monod, J., Wyman, J., and Changeux, J.-P., 1965. On the nature of allosteric transitions: A plausible model. Journal of Molecular Biology 12:92.*)

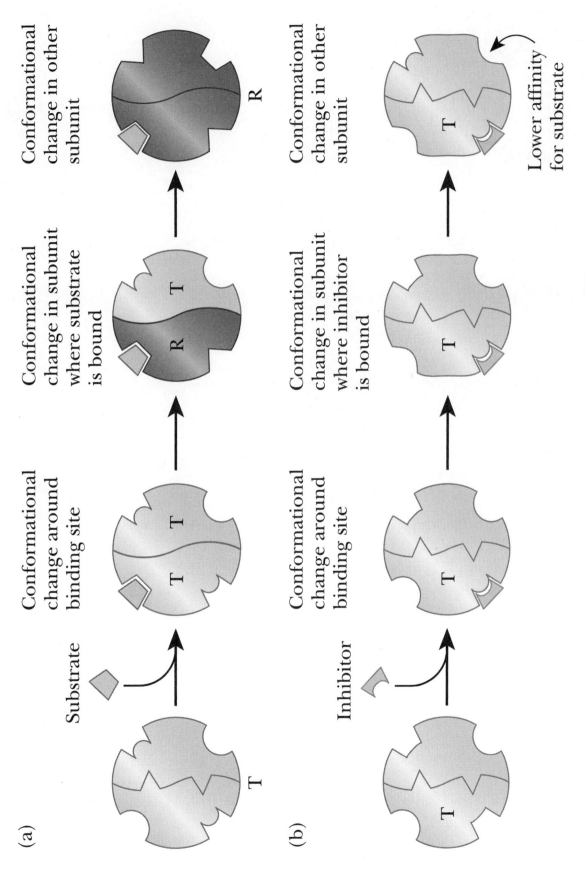

Figure 7.7 Sequential model of cooperative binding of substrate to allosteric enzyme

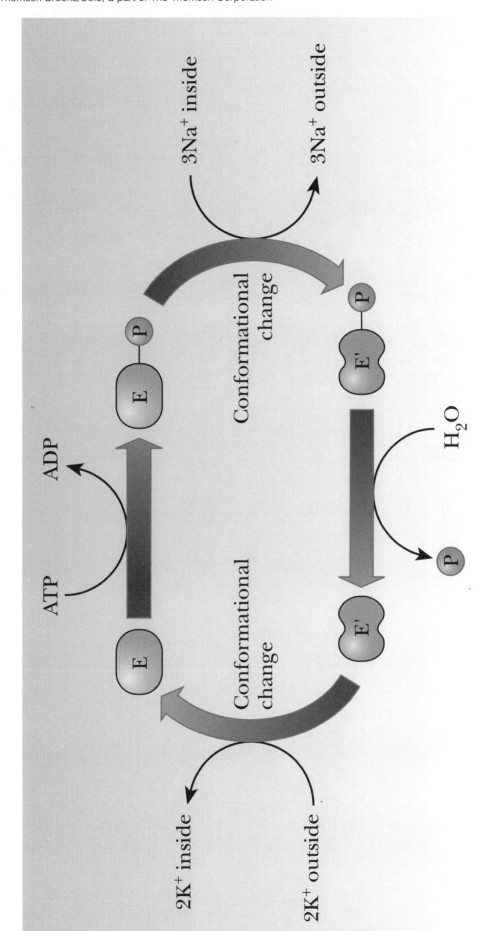

Figure 7.8 Phosphorylation of the sodium-potassium pump

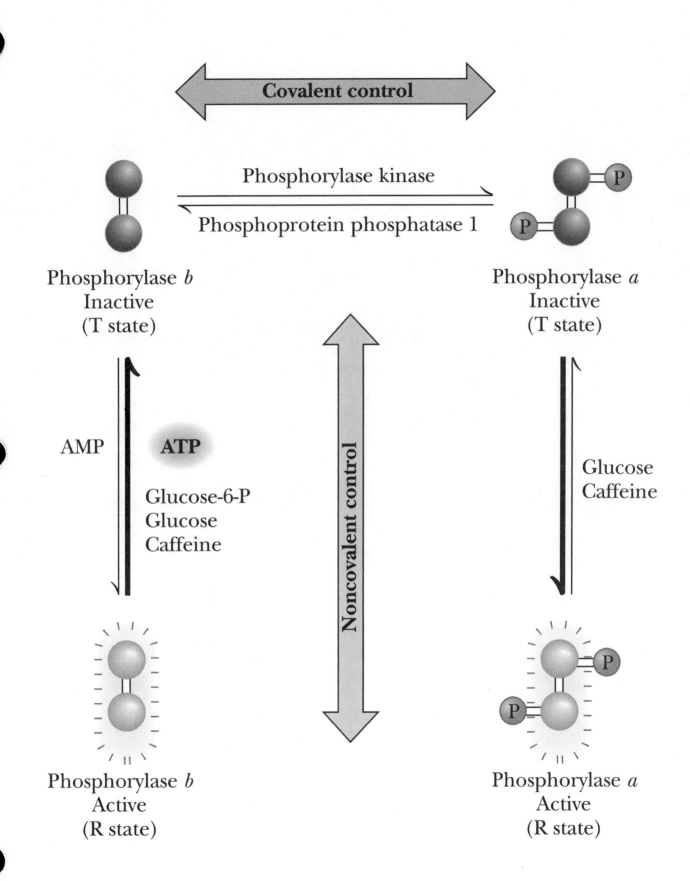

Figure 7.9 Glycogen phosphorylation activity

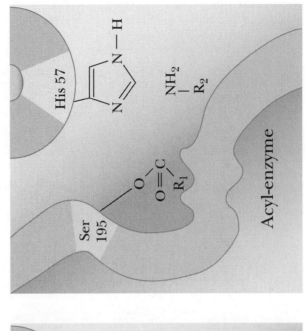

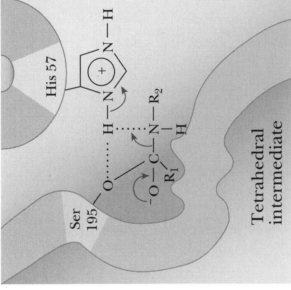

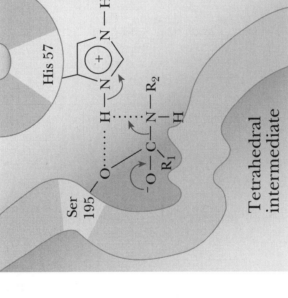

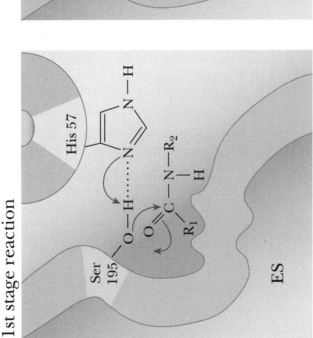

33

1st stage reaction

ES

Tetrahedral
intermediate

Acyl-enzyme

His 57

Ser
195

(*From Hammes, G.: Enzyme Catalysis and Regulation, New York: Academic Press, 1982.*)

Figure 7.14 (top) The mechanism of chymotrypsin action

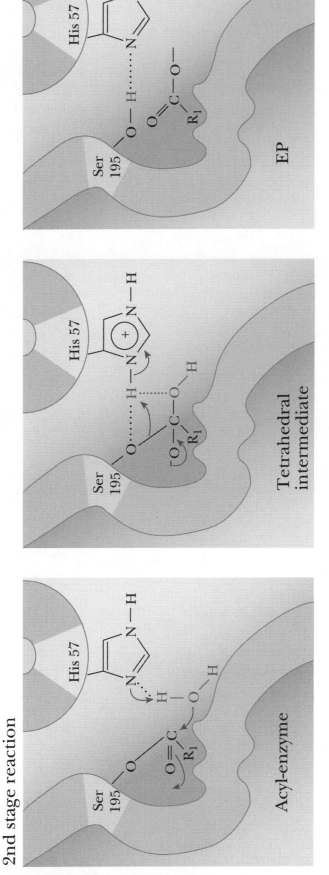

2nd stage reaction

Acyl-enzyme

Tetrahedral intermediate

EP

(*From Hammes, G.: Enzyme Catalysis and Regulation, New York: Academic Press, 1982.*)

Figure 7.14 (bottom) *The mechanism of chymotrypsin action*

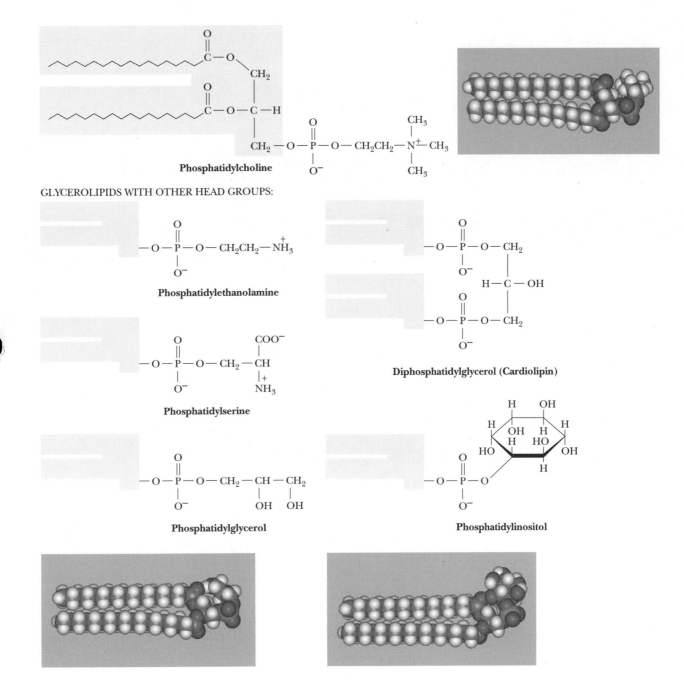

35

Phosphatidylcholine

GLYCEROLIPIDS WITH OTHER HEAD GROUPS:

Phosphatidylethanolamine

Phosphatidylserine

Phosphatidylglycerol

Diphosphatidylglycerol (Cardiolipin)

Phosphatidylinositol

Figure 8.5 Structures of some phosphoacylglycerols

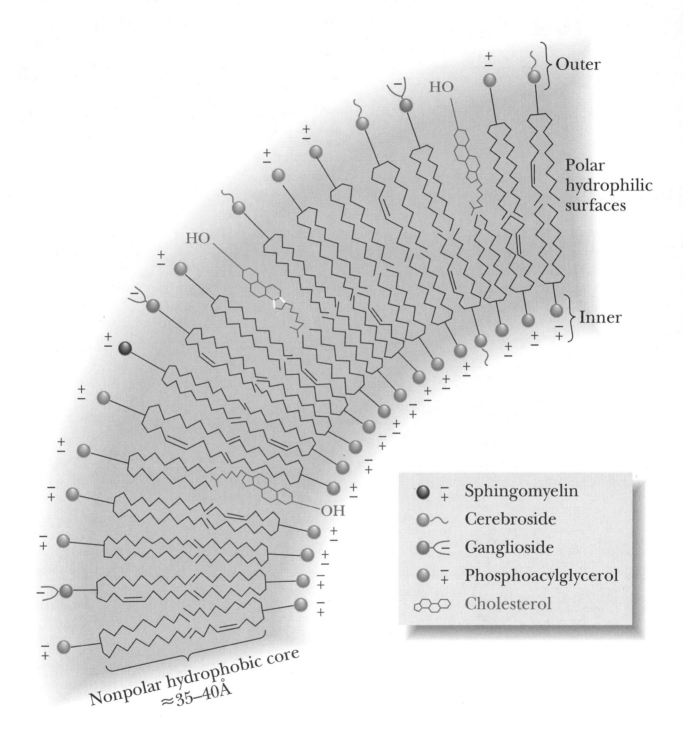

Outer

Polar
hydrophilic
surfaces

HO

Inner

HO

OH

Nonpolar hydrophobic core
≈35–40Å

⬤ $\frac{-}{+}$	Sphingomyelin
⬤∿	Cerebroside
⬤⪇	Ganglioside
⬤ $\frac{-}{+}$	Phosphoacylglycerol
⬡⬡	Cholesterol

Figure 8.11 Lipid bilayer asymmetry

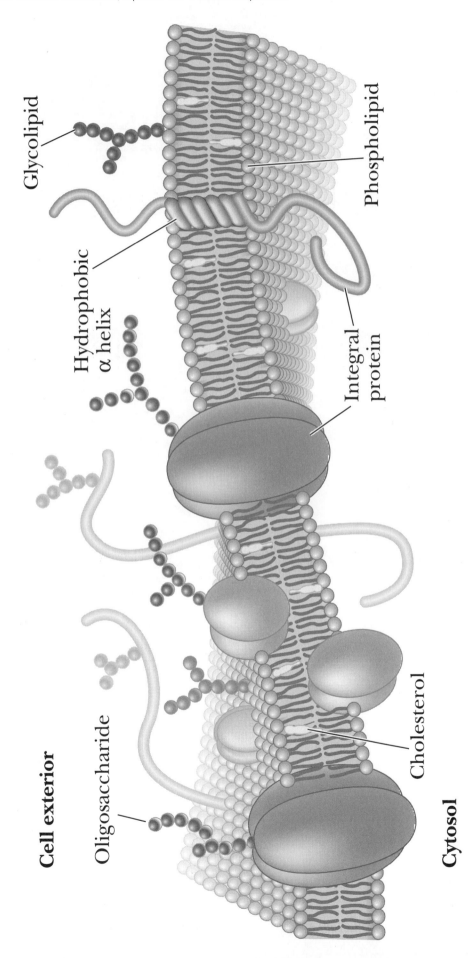

37

Figure 8.18 Fluid-mosaic model for membrane structure

(From Singer, S. J., in G. Weissman and R. Claiborne, Eds., Cell Membranes: Biochemistry, Cell Biology, and Pathology, New York: HP Pub., 1975, p. 37.)

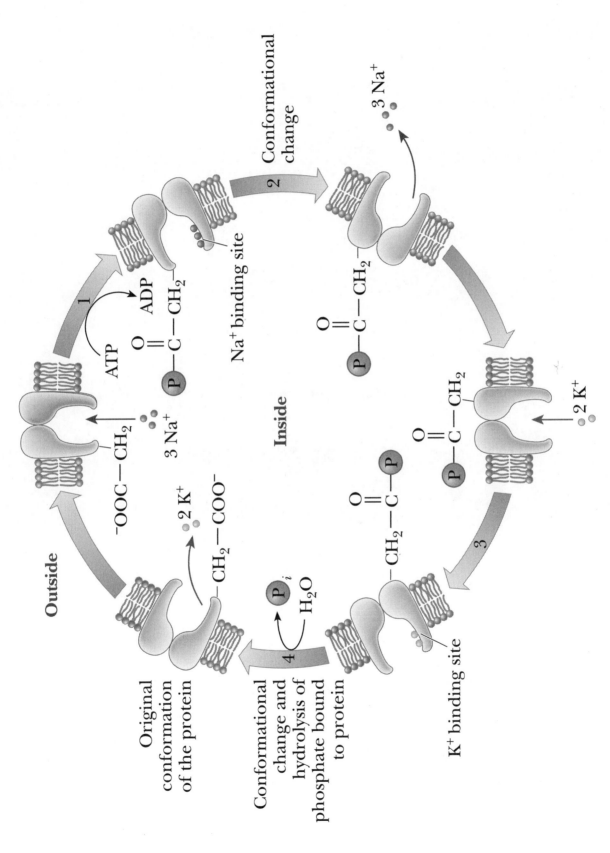

Figure 8.24 The sodium-potassium ion pump

(a)

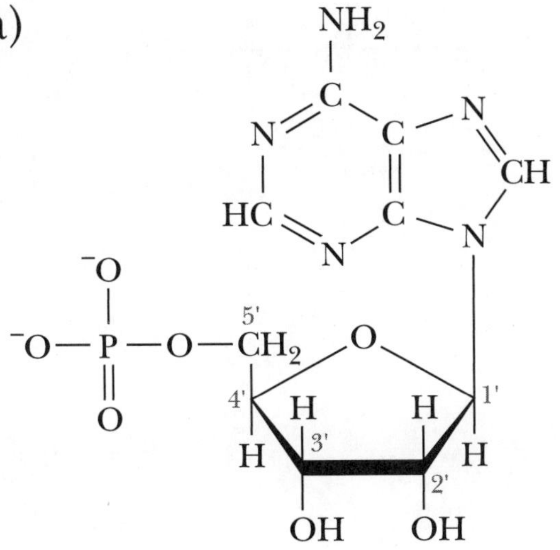

Adenosine 5'-monophosphate

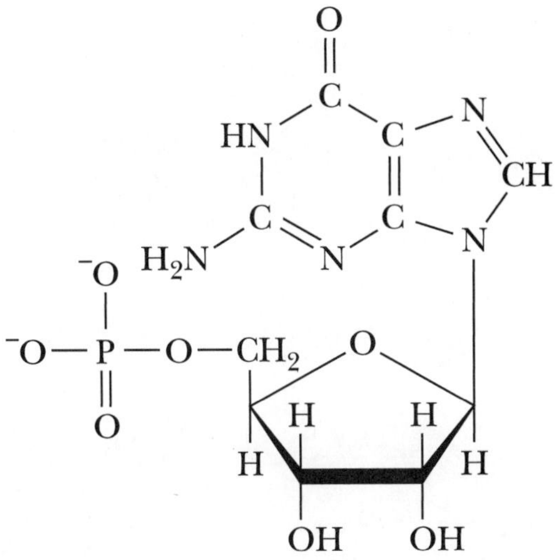

Guanosine 5'-monophosphate

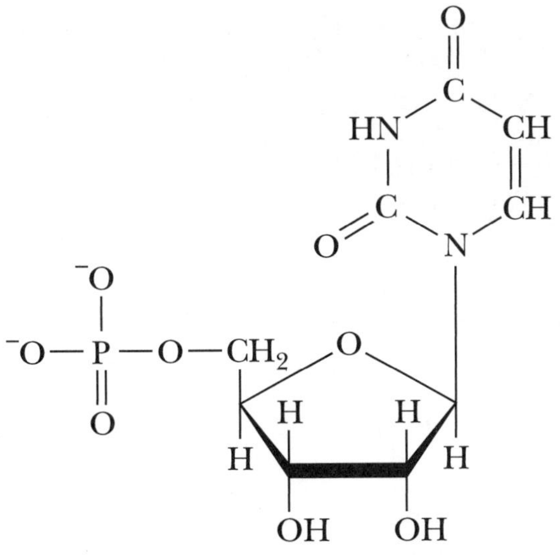

Uridine 5'-monophosphate

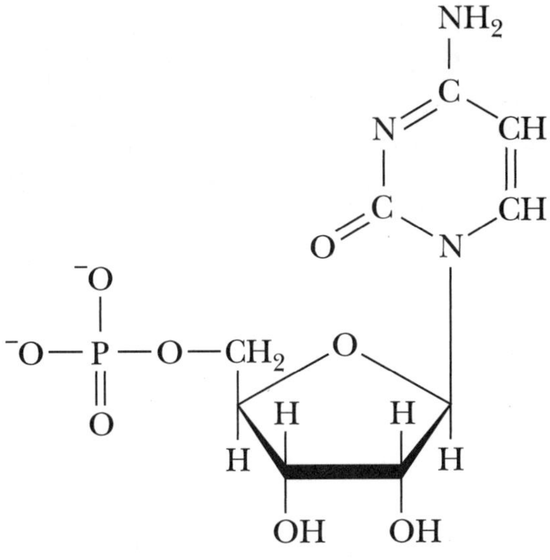

Cytidine 5'-monophosphate

Figure 9.4a Commonly occurring nucleotides

(b)

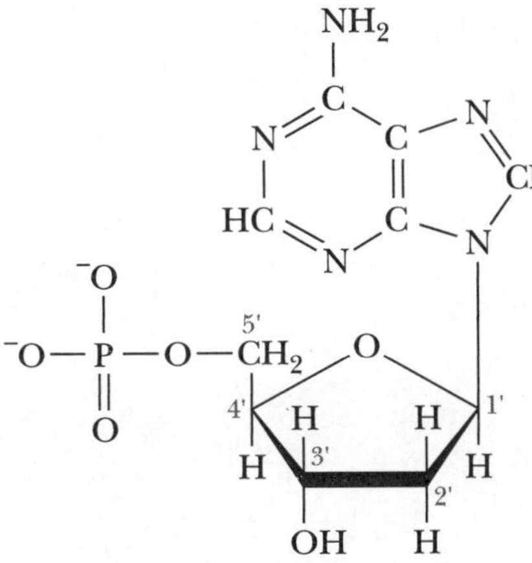

Deoxyadenosine 5'-monophosphate

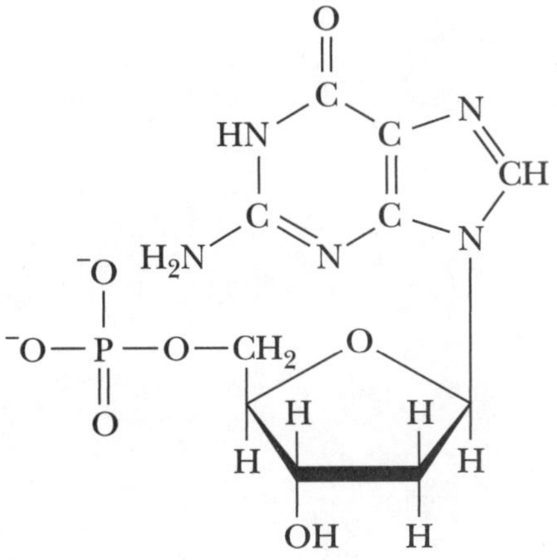

Deoxyguanosine 5'-monophosphate

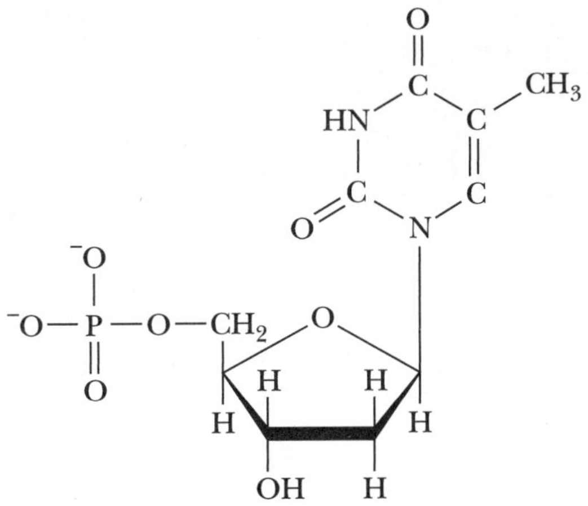

Deoxythymidine 5'-monophosphate

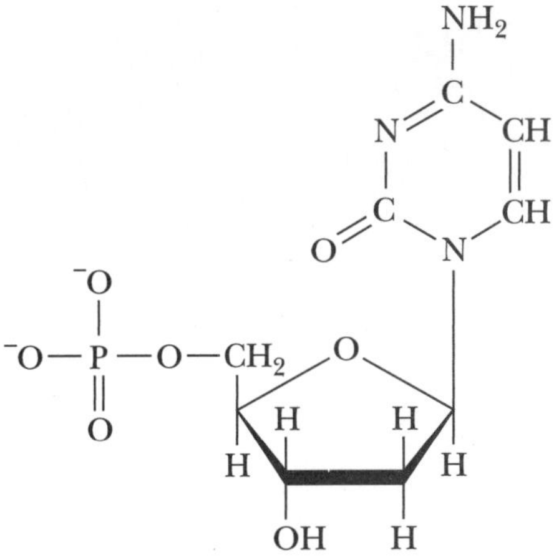

Deoxycytidine 5'-monophosphate

Figure 9.4b Commonly occurring nucleotides

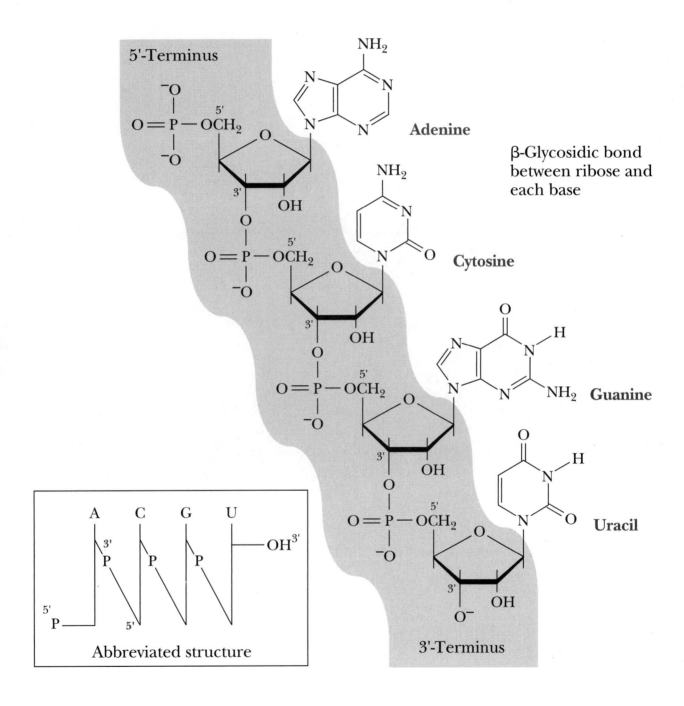

Figure 9.5 A fragment of an RNA chain

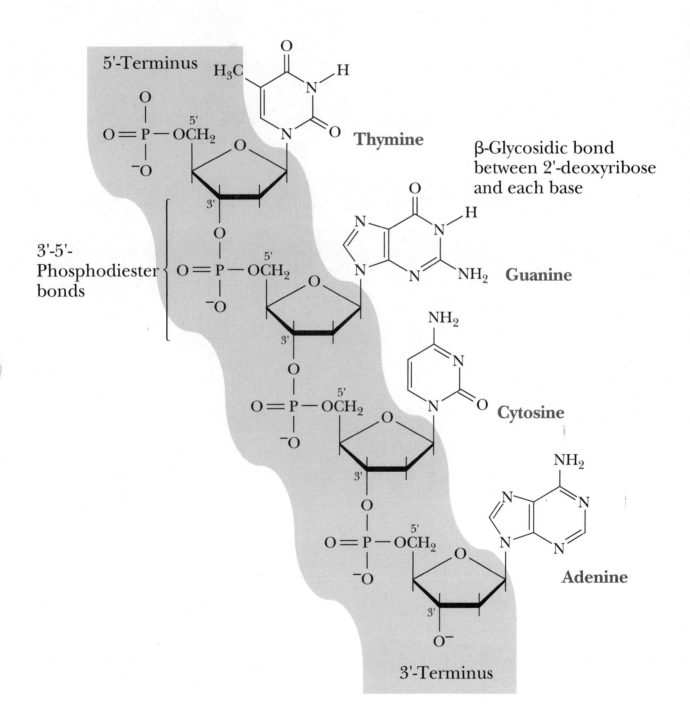

Figure 9.6 A portion of a DNA chain

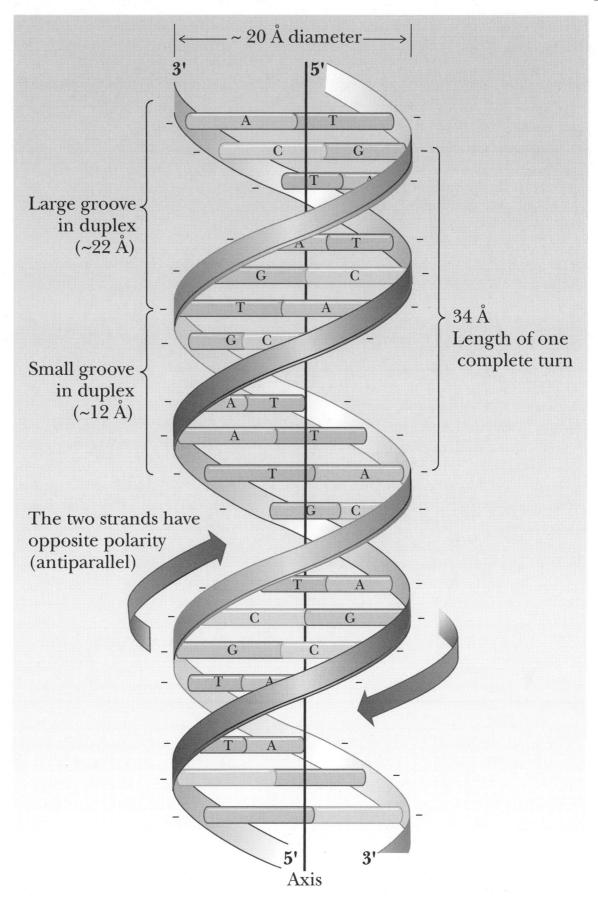

Figure 9.7 The double helix

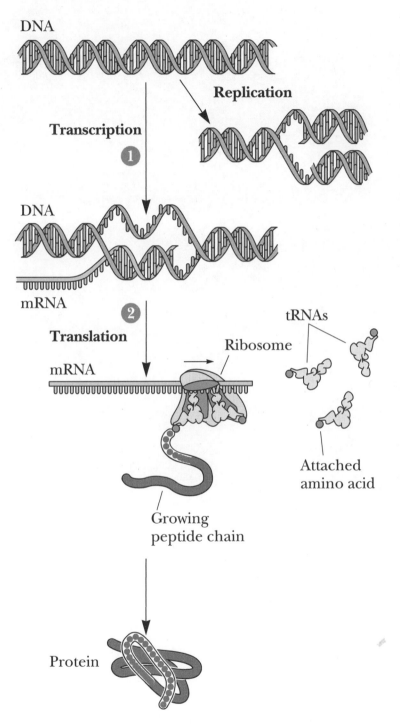

Replication

DNA replication yields two DNA molecules identical to the original one, ensuring transmission of genetic information to daughter cells with exceptional fidelity.

Transcription

The sequence of bases in DNA is recorded as a sequence of complementary bases in a single-stranded mRNA molecule.

Translation

Three-base codons on the mRNA corresponding to specific amino acids direct the sequence of building a protein. These codons are recognized by tRNAs (transfer RNAs) carrying the appropriate amino acids. Ribosomes are the "machinery" for protein synthesis.

Figure 9.19 Types of RNA

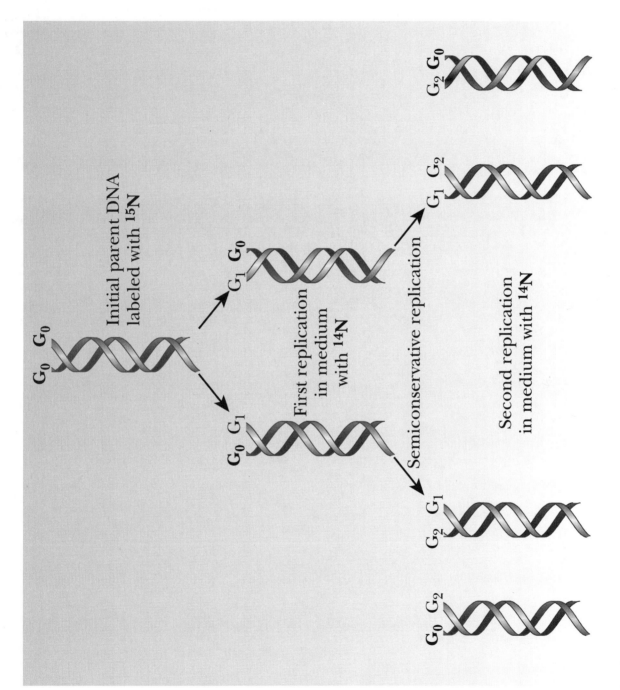

Figure 10.2 Semiconservative replication

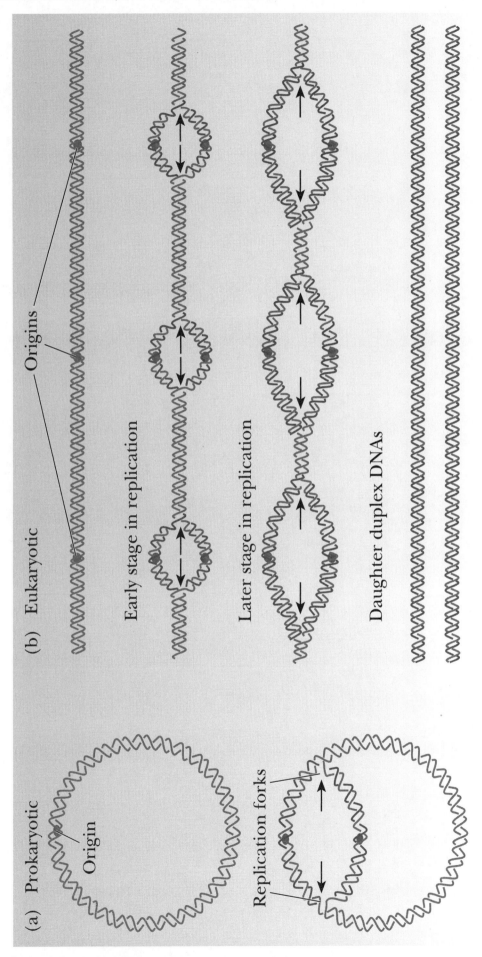

Figure 10.4 *Bidirectional replication of DNA in prokaryotes and eukaryotes*

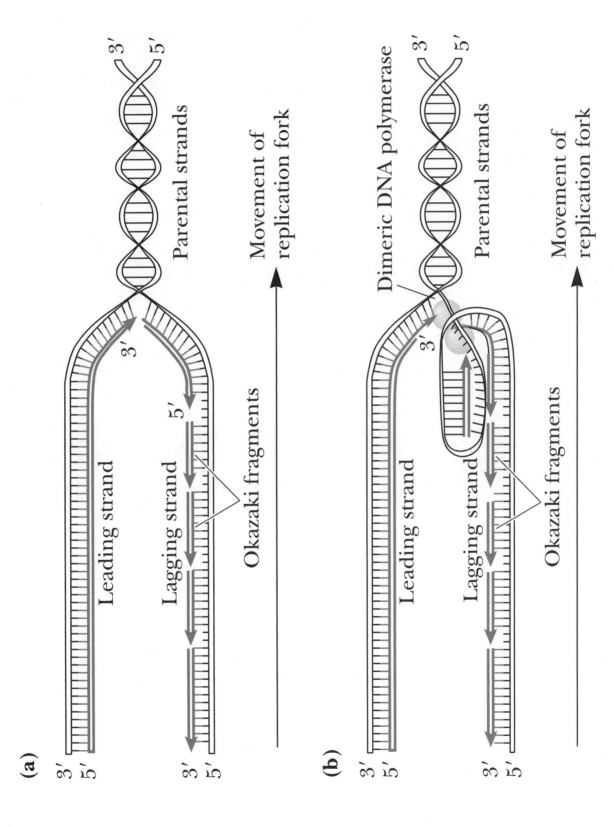

Figure 10.5 The semidiscontinuous model for DNA replication

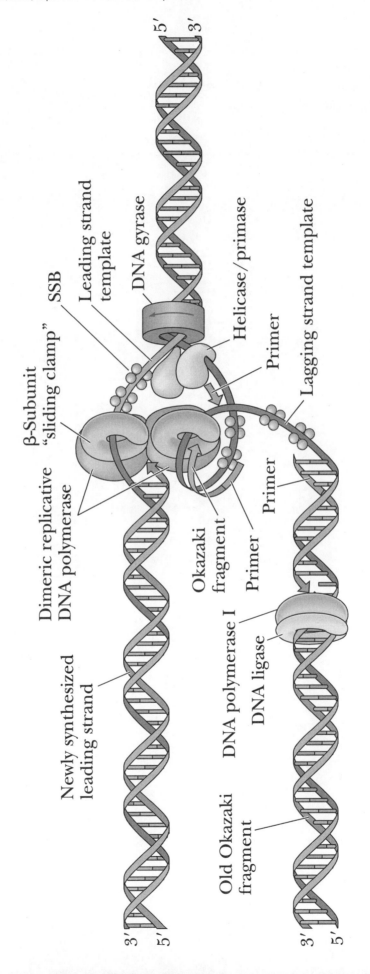

Figure 10.10 General features of the replication fork

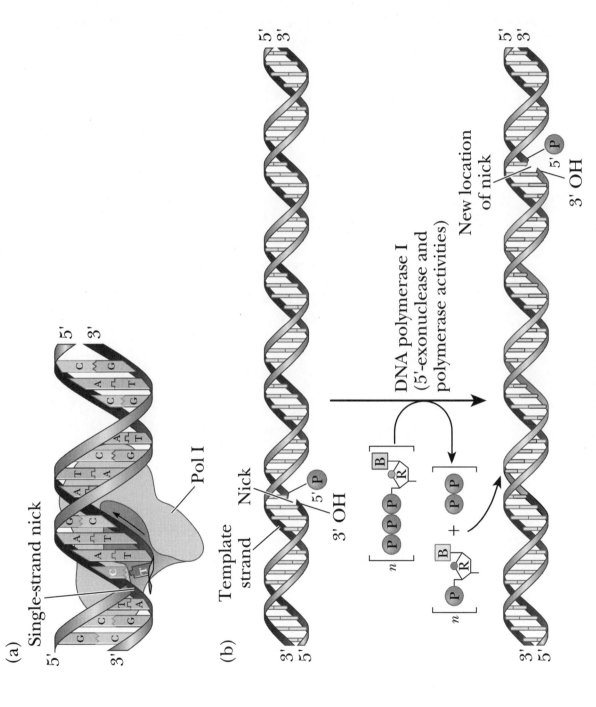

Figure 10.12 The 5′→ 3′ exonuclease activity of DNA polymerase I

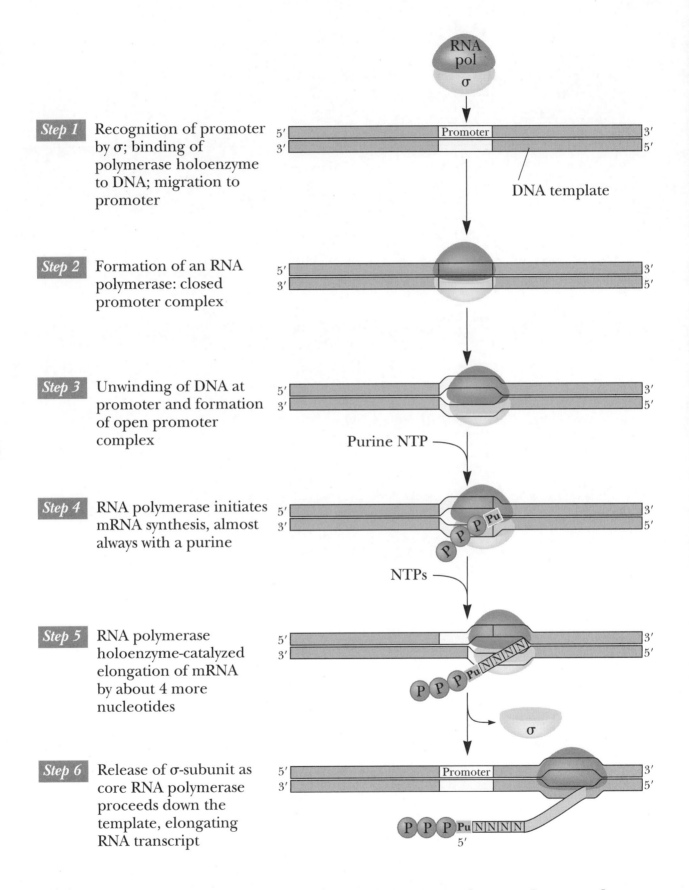

Step 1 Recognition of promoter by σ; binding of polymerase holoenzyme to DNA; migration to promoter

Step 2 Formation of an RNA polymerase: closed promoter complex

Step 3 Unwinding of DNA at promoter and formation of open promoter complex

Step 4 RNA polymerase initiates mRNA synthesis, almost always with a purine

Step 5 RNA polymerase holoenzyme-catalyzed elongation of mRNA by about 4 more nucleotides

Step 6 Release of σ-subunit as core RNA polymerase proceeds down the template, elongating RNA transcript

Figure 11.3 The basic order of events in prokaryotic transcription initiation and elongation

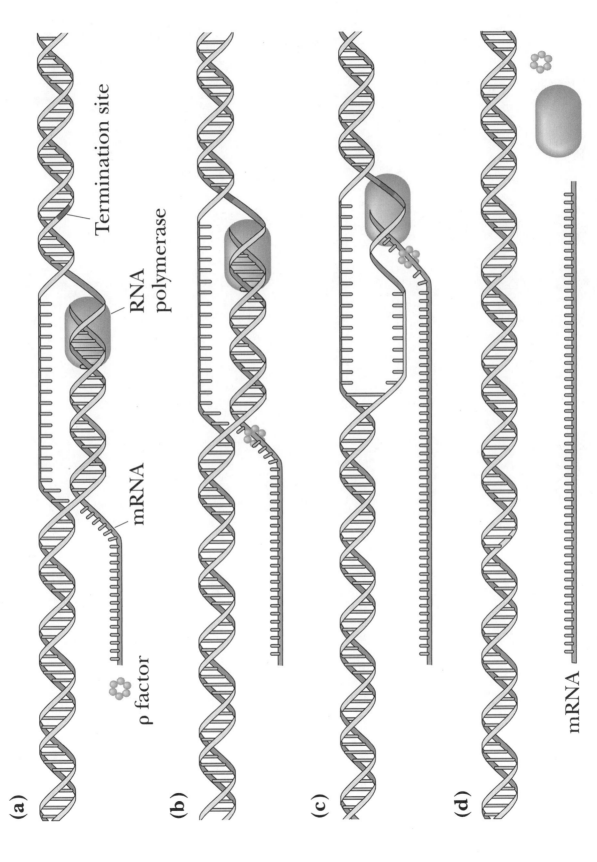

Figure 11.6 Transcription termination by the rho factor

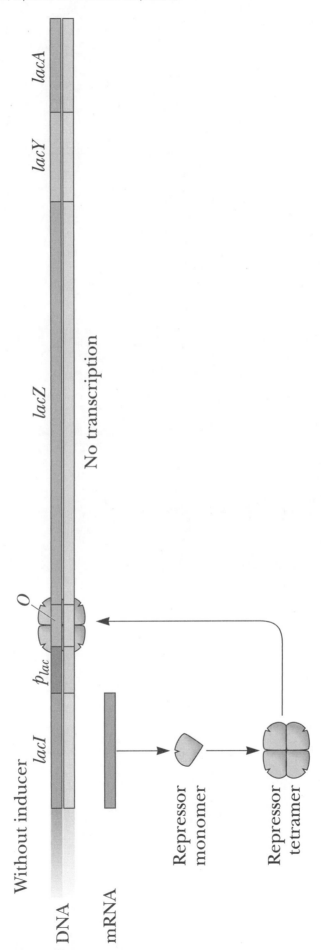

Figure 11.9 (top) **The lacI gene produces a protein that represses the lac operon**

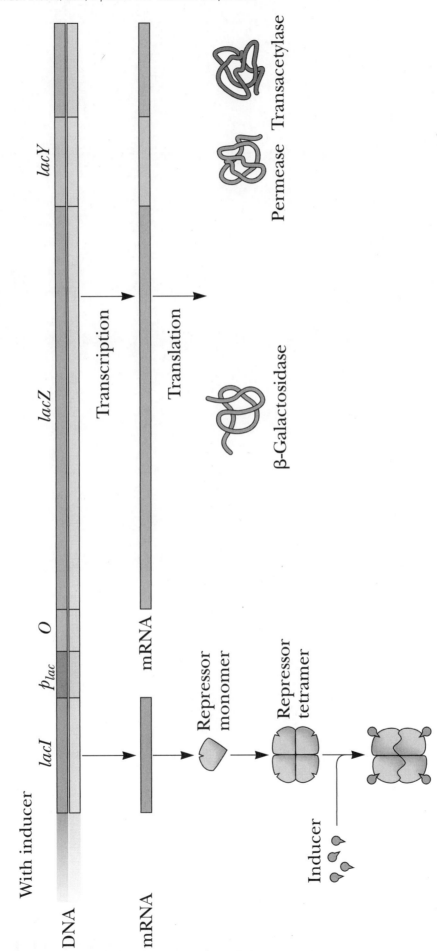

Figure 11.9 (bottom) *The lacI gene produces a protein that represses the lac operon*

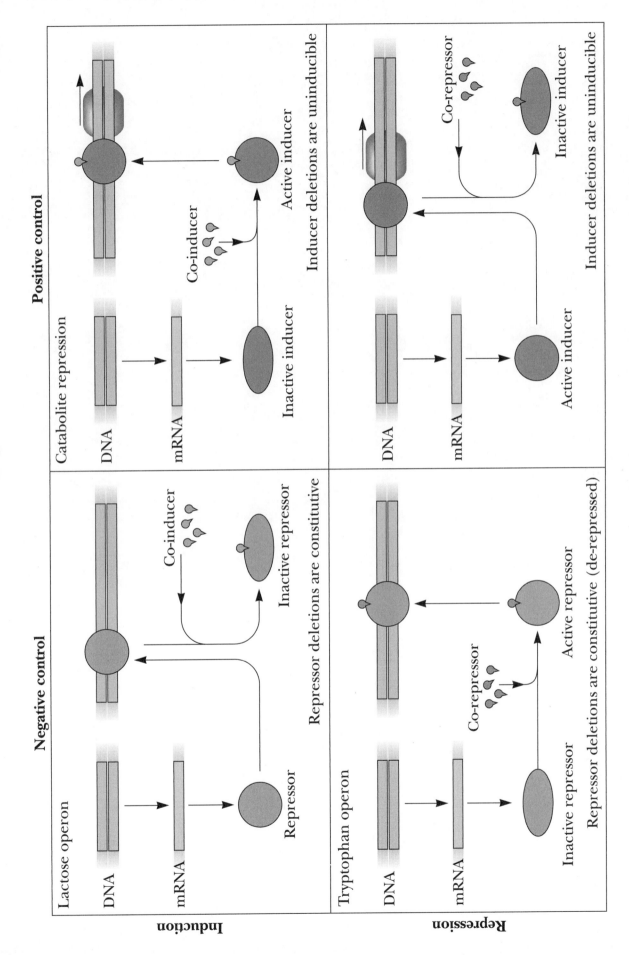

Figure 11.12 **Basic control mechanisms seen in the expression of genes**

55

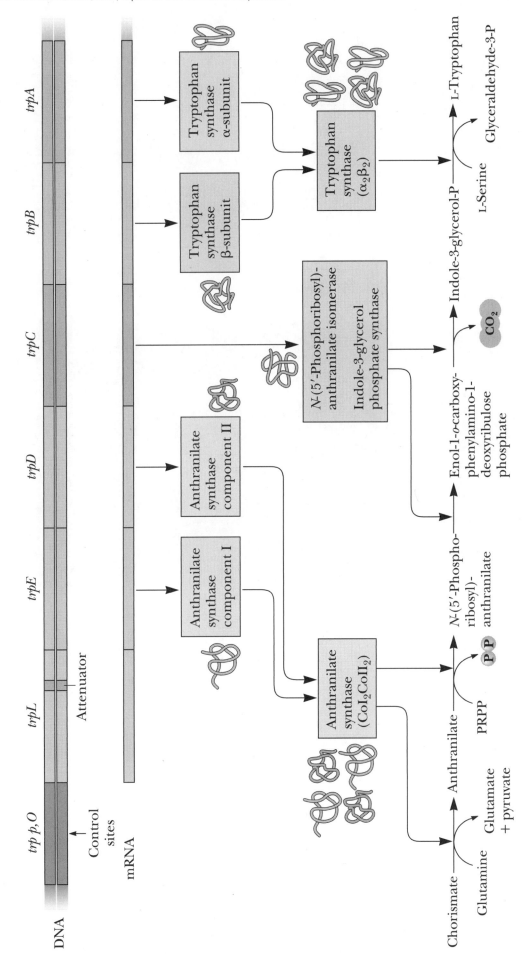

Figure 11.13 *The trp operon of E. coli*

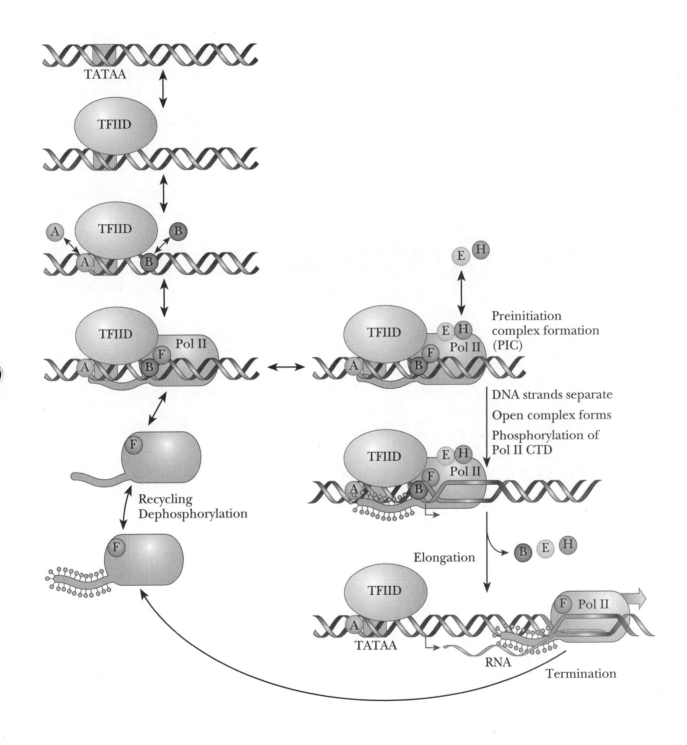

Figure 11.18 The order of events of transcription

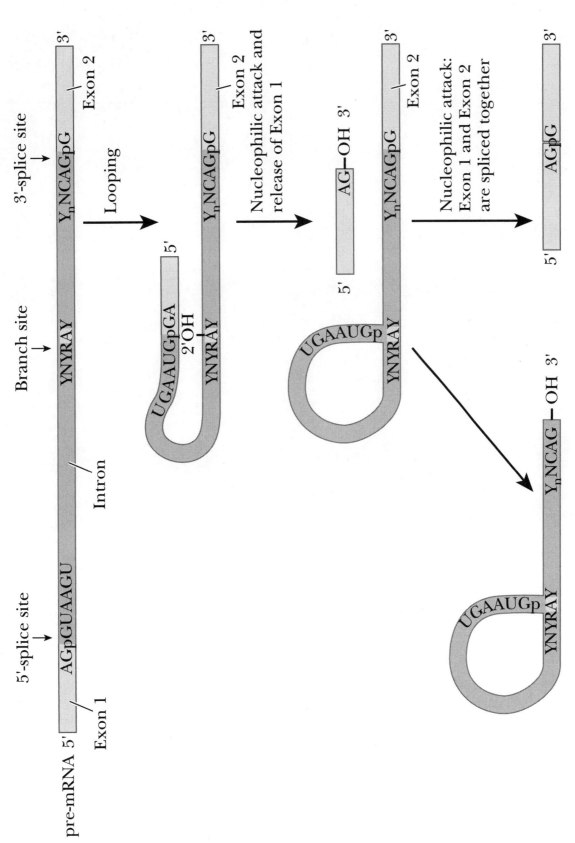

Figure 11.34 Splicing of mRNA precursors

(Adapted from Sharp, P. A., 1987, Science **235**, 766, Figure 1.)

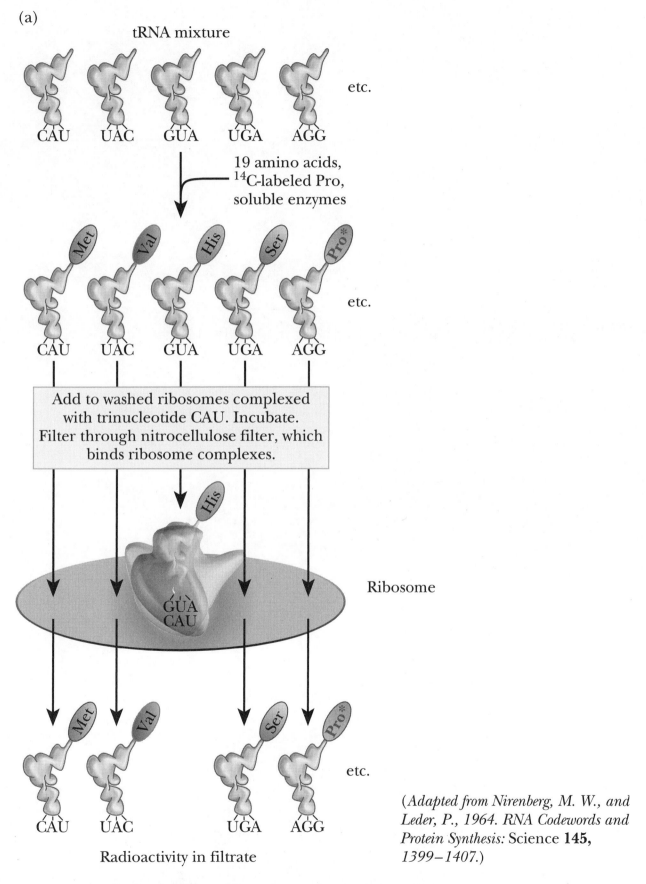

Figure 12.3a The filter-binding assay for elucidation of the genetic code

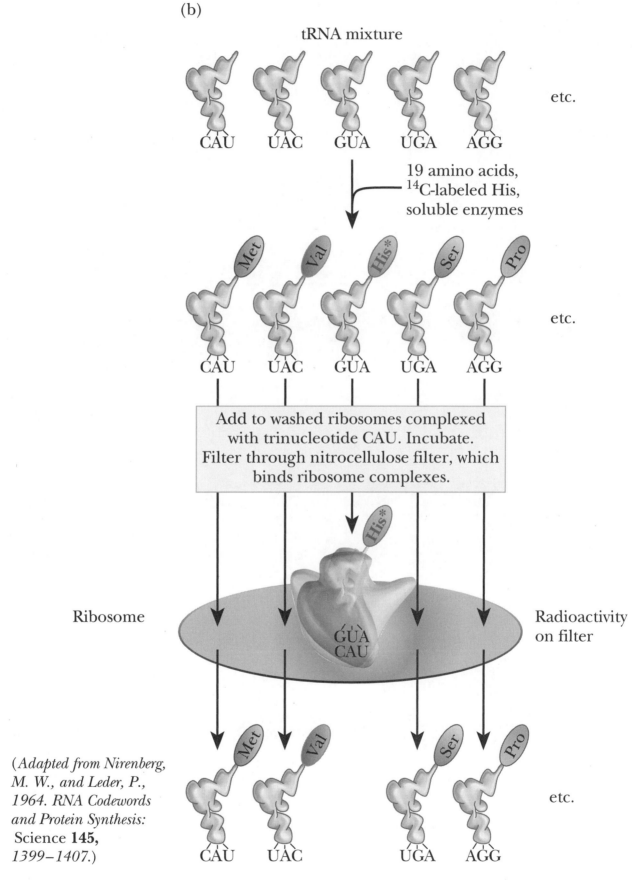

Figure 12.3b The filter-binding assay for elucidation of the genetic code

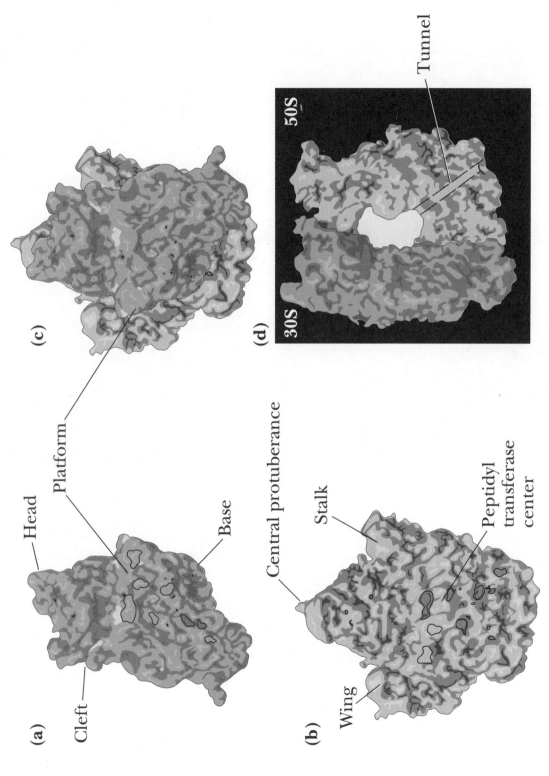

(a)

Head

Platform

Cleft

Base

(b)

Central protuberance

Stalk

Wing

Peptidyl transferase center

(c)

(d)

30S

50S

Tunnel

(Adapted from Figures 2 and 3 in Cate, J. H., et al., 1999. X-ray crystal structures of 70S ribosomal functional complexes. Science **285**, 2095–2104.)

Figure 12.8 Structure of ribosome

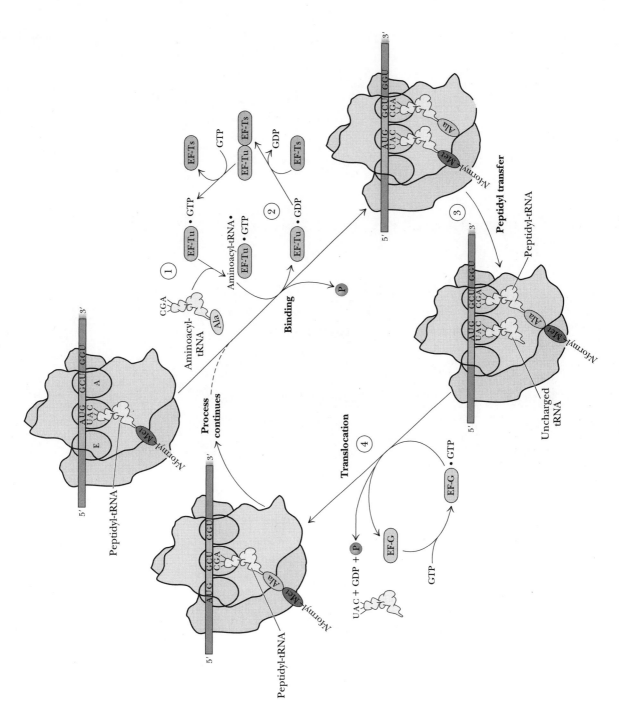

Figure 12.12 A summary of the steps in chain elongation

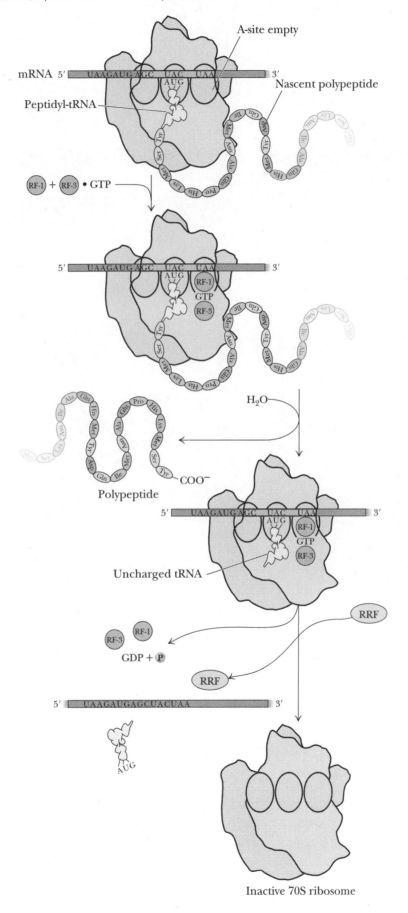

Figure 12.15 **The events in peptide chain termination**

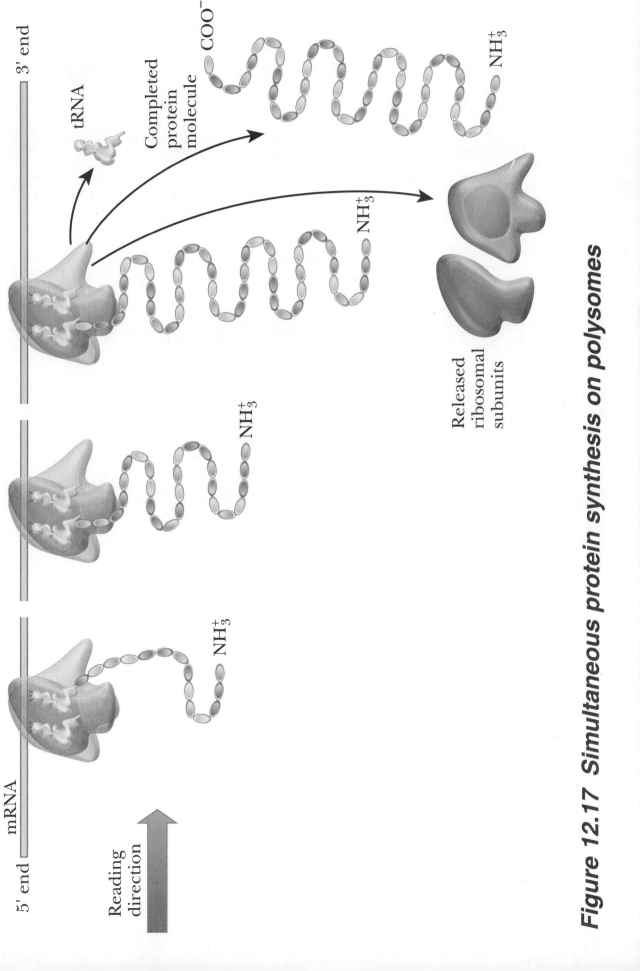

Figure 12.17 Simultaneous protein synthesis on polysomes

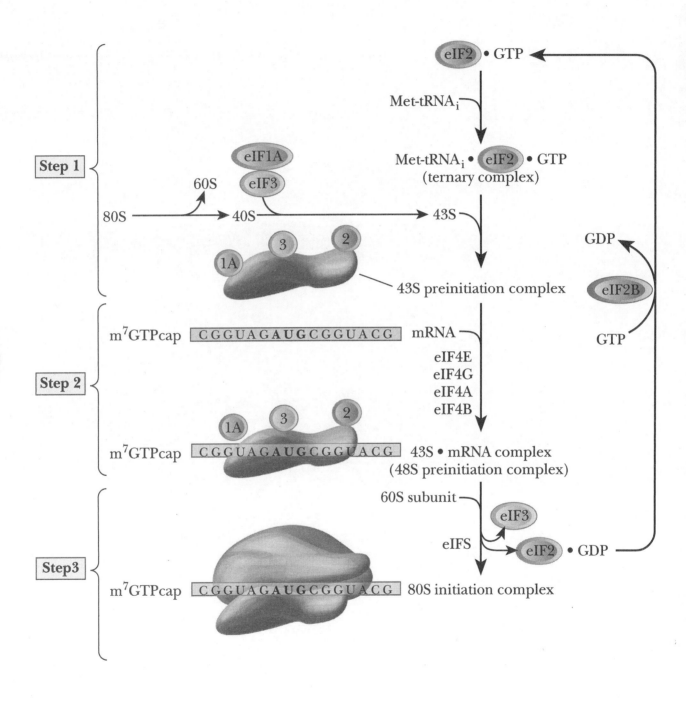

Figure 12.20 The three stages in the initiation of translation in eukaryotic cells

Table 12.1

The Genetic Code

First Position (5'-end)	Second Position				Third Position (3'-end)
	U	C	A	G	
U	UUU Phe	UCU Ser	UAU Tyr	UGU Cys	U
	UUC Phe	UCC Ser	UAC Tyr	UGC Cys	C
	UUA Leu	UCA Ser	UAA Stop	UGA Stop	A
	UUG Leu	UCG Ser	UAG Stop	UGG Trp	G
C	CUU Leu	CCU Pro	CAU His	CGU Arg	U
	CUC Leu	CCC Pro	CAC His	CGC Arg	C
	CUA Leu	CCA Pro	CAA Gln	CGA Arg	A
	CUG Leu	CCG Pro	CAG Gln	CGG Arg	G
A	AUU Ile	ACU Thr	AAU Asn	AGU Ser	U
	AUC Ile	ACC Thr	AAC Asn	AGC Ser	C
	AUA Ile	ACA Thr	AAA Lys	AGA Arg	A
	AUG Met*	ACG Thr	AAG Lys	AGG Arg	G
G	GUU Val	GCU Ala	GAU Asp	GGU Gly	U
	GUC Val	GCC Ala	GAC Asp	GGC Gly	C
	GUA Val	GCA Ala	GAA Glu	GGA Gly	A
	GUG Val	GCG Ala	GAG Glu	GGG Gly	G

Third-Base Degeneracy Is Color-Coded

Third-Base Relationship	Third Bases with Same Meaning	Number of Codons
Third-base irrelevant	U, C, A, G	32 (8 families)
Purines	A or G	12 (6 pairs)
Pyrimidines	U or C	14 (7 pairs)
Three out of four	U, C, A	3 (AUX = Ile)
Unique definitions	G only	2 (AUG = Met) (UGG = Trp)
Unique definition	A only	1 (UGA = Stop)

*AUG signals translation initiation as well as coding for Met residues.

Table 12.1 The genetic code

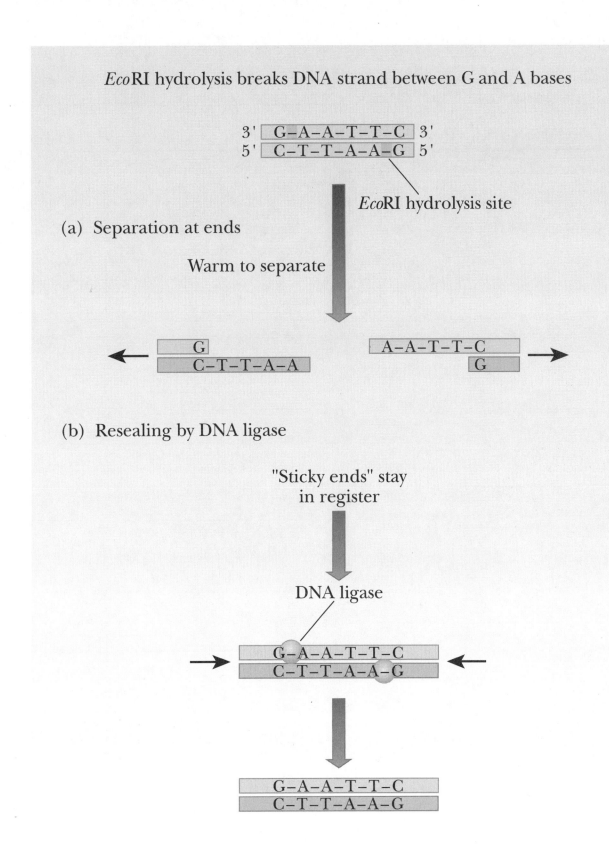

Figure 13.4 Hydrolysis of DNA by restriction endonucleases

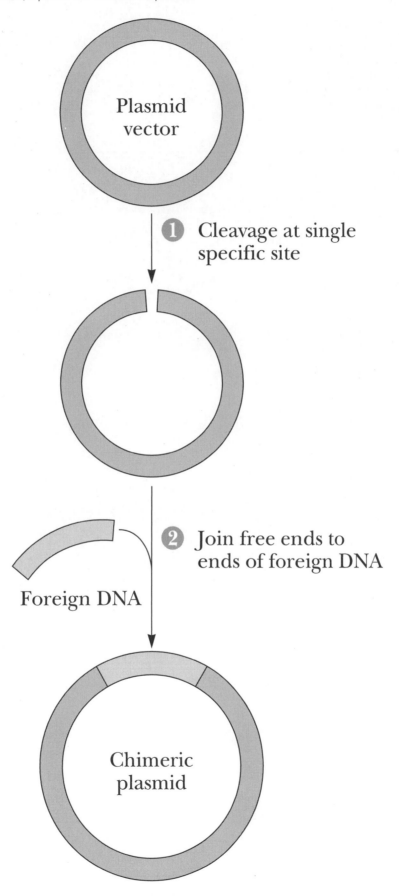

Figure 13.5 Cloning of foreign DNA into a bacterial plasmid

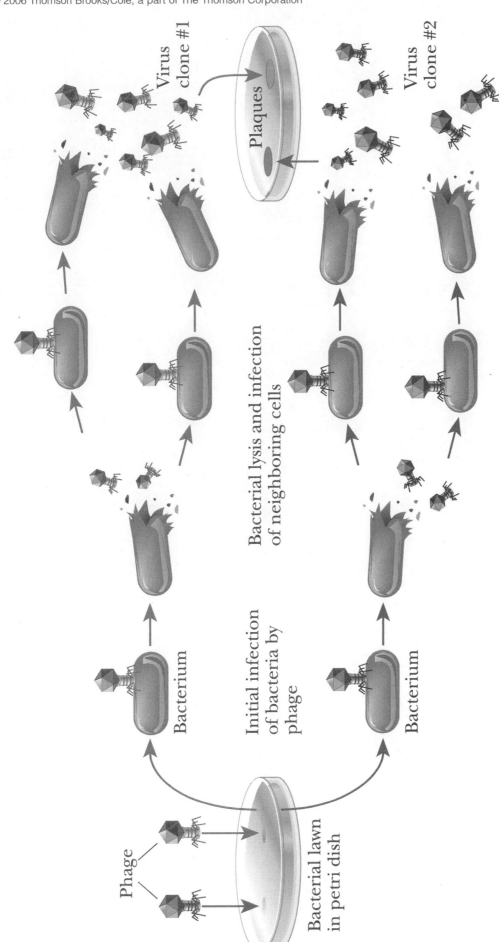

Virus clone #1

Virus clone #2

Plaques

Bacterial lysis and infection of neighboring cells

Bacterium

Bacterium

Initial infection of bacteria by phage

Phage

Bacterial lawn in petri dish

(Adapted with permission from Dealing with Genes: The Language of Heredity, by Paul Berg and Maxine Singer, © 1992 by University Science Books.)

Figure 13.6 The cloning of a virus

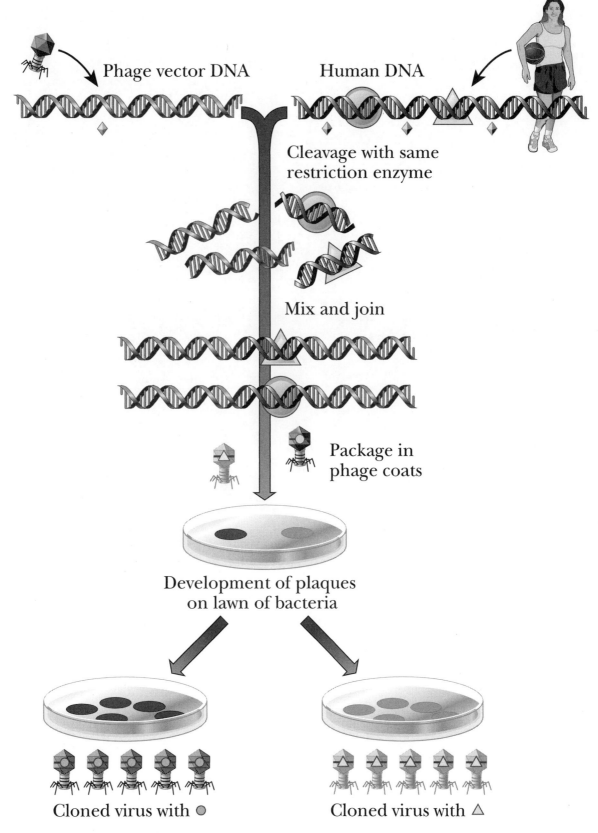

Figure 13.8 The cloning of human DNA fragments with a viral vector

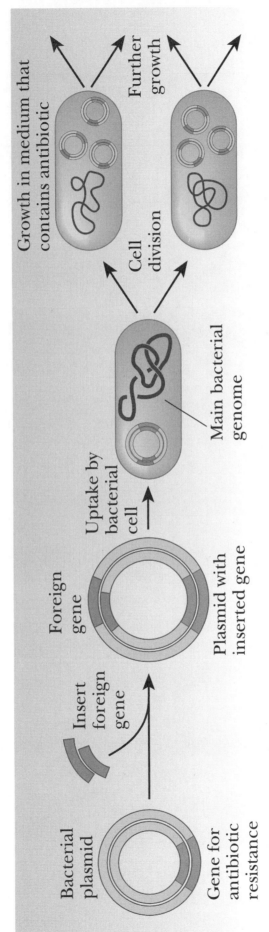

70

(*Adapted with permission from Dealing with Genes: The Language of Heredity, by Paul Berg and Maxine Singer,*
© 1992 by University Science Books.)

Figure 13.9 *Selecting for recombinant DNA in a bacterial plasmid*

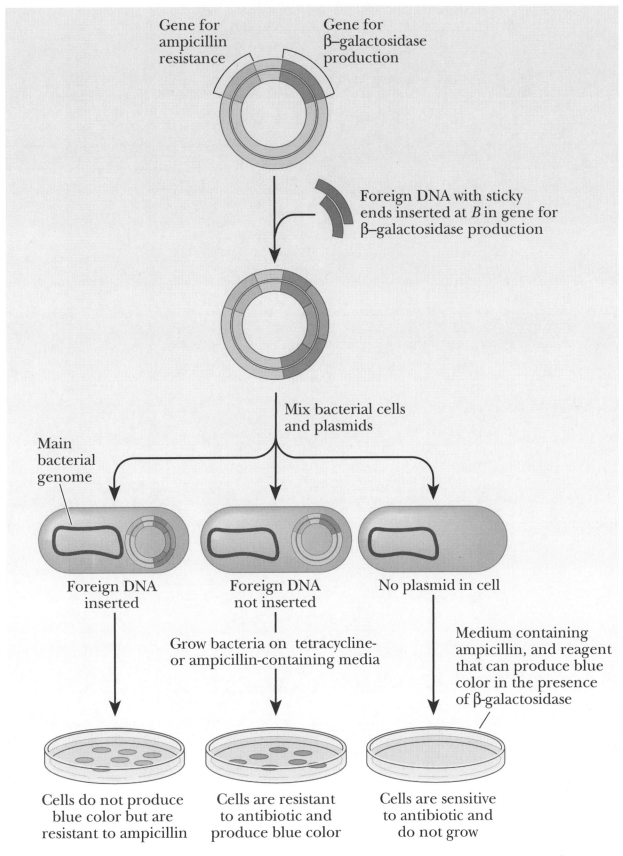

(*Adapted with permission from* Dealing with Genes: The Language of Heredity, *by Paul Berg and Maxine Singer, © 1992 by University Science Books.*)

Figure 13.13 Clone selection via blue/white screening

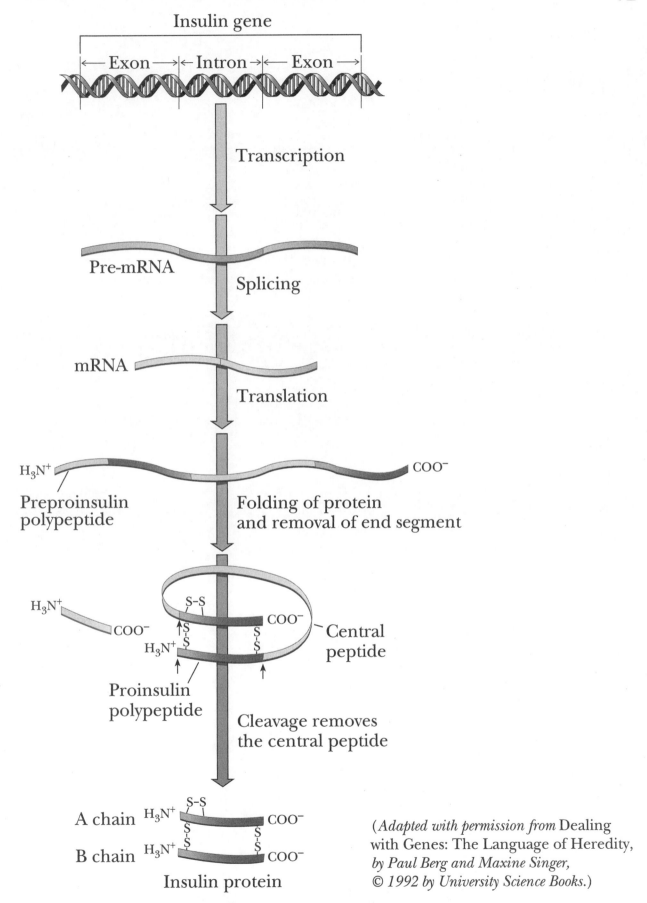

Figure 13.15 Synthesis of insulin in humans

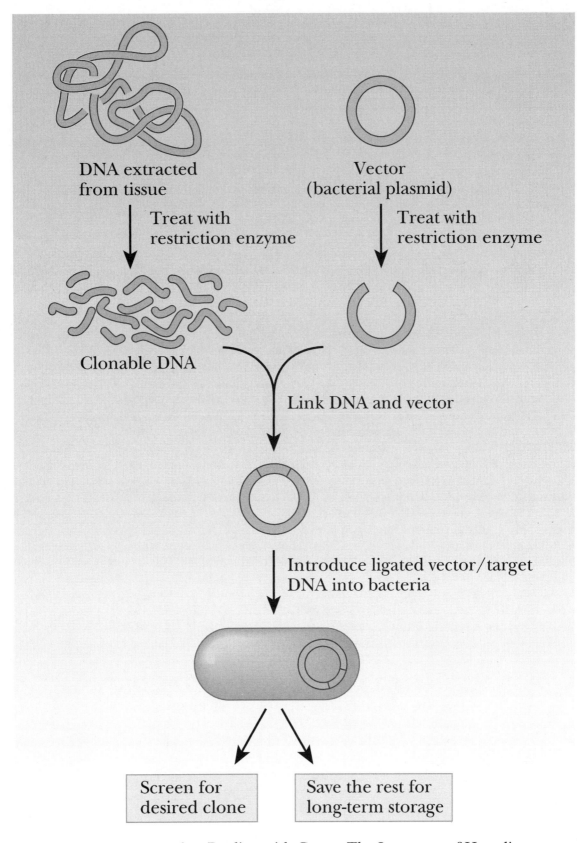

73

(*Adapted with permission from* Dealing with Genes: The Language of Heredity, *by Paul Berg and Maxine Singer,* © 1992 by University Science Books.)

Figure 13.19 Steps involved in the construction of a DNA library

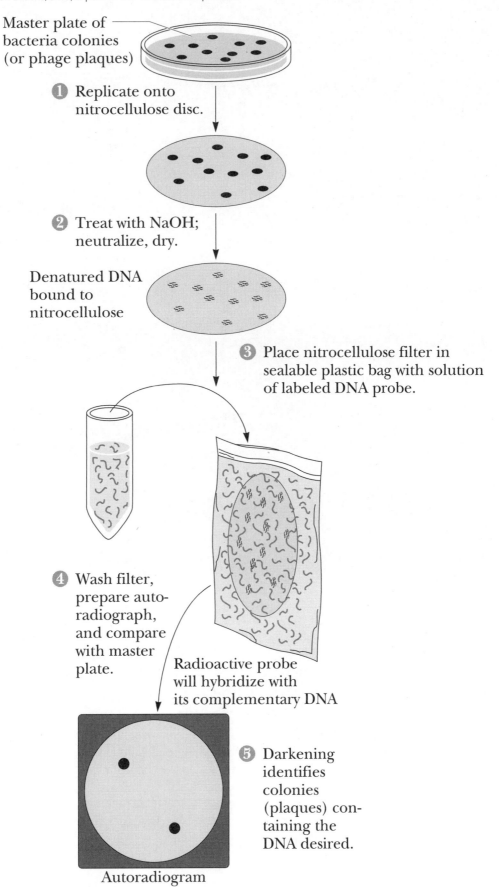

Master plate of
bacteria colonies
(or phage plaques)

1 Replicate onto
nitrocellulose disc.

2 Treat with NaOH;
neutralize, dry.

Denatured DNA
bound to
nitrocellulose

3 Place nitrocellulose filter in
sealable plastic bag with solution
of labeled DNA probe.

4 Wash filter,
prepare auto-
radiograph,
and compare
with master
plate.

Radioactive probe
will hybridize with
its complementary DNA

5 Darkening
identifies
colonies
(plaques) con-
taining the
DNA desired.

Autoradiogram

*Figure 13.20 Selecting a desired clone from a DNA
library*

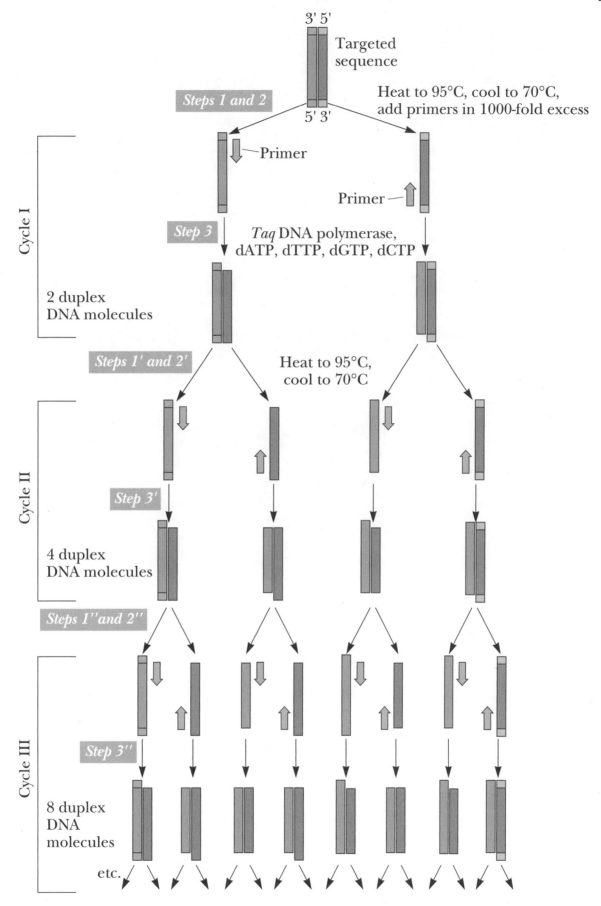

Figure 13.22 The polymerase chain reaction

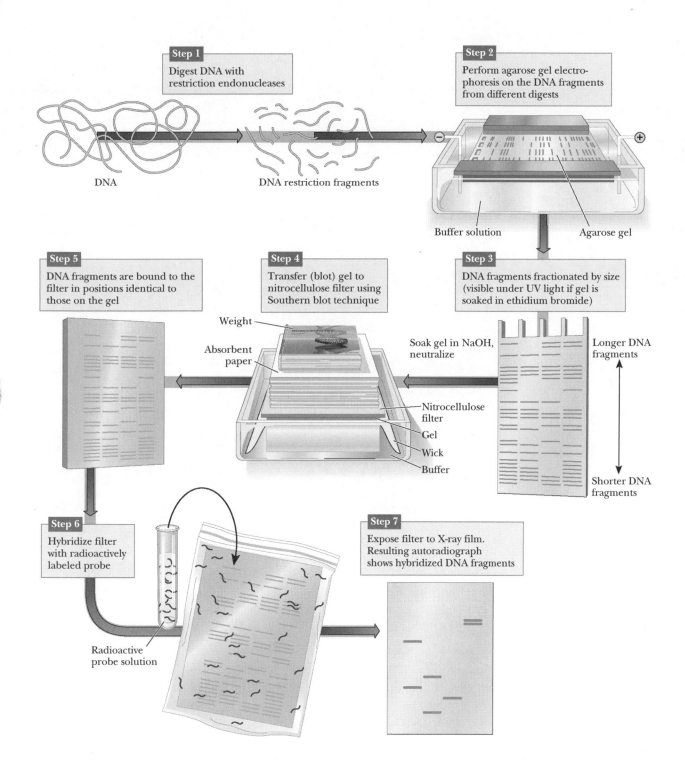

Figure 13.23 The Southern blot

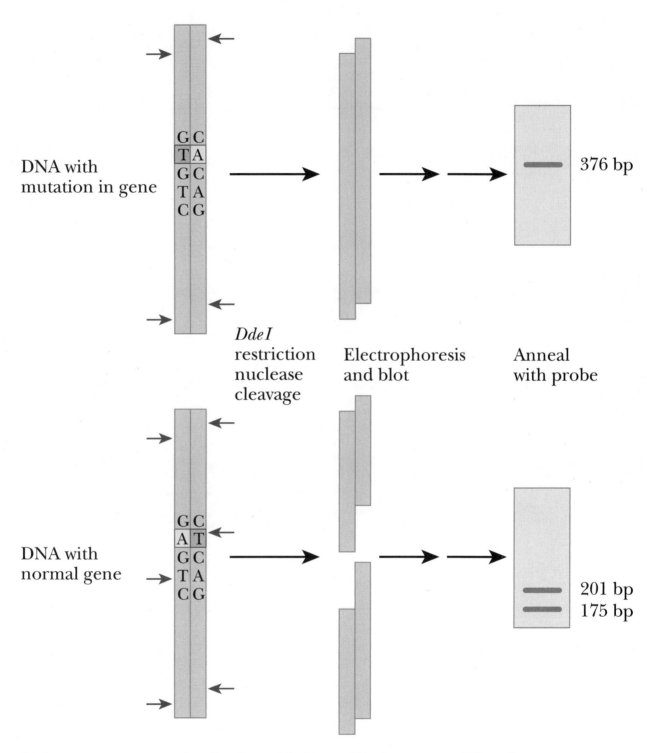

DNA with
mutation in gene

Dde I
restriction
nuclease
cleavage

Electrophoresis
and blot

Anneal
with probe

376 bp

DNA with
normal gene

201 bp
175 bp

(*Adapted with permission from* Dealing with Genes: The Language of Heredity,
by Paul Berg and Maxine Singer, © 1992 by University Science Books.)

Figure 13.24 Restriction-fragment length polymorphisms

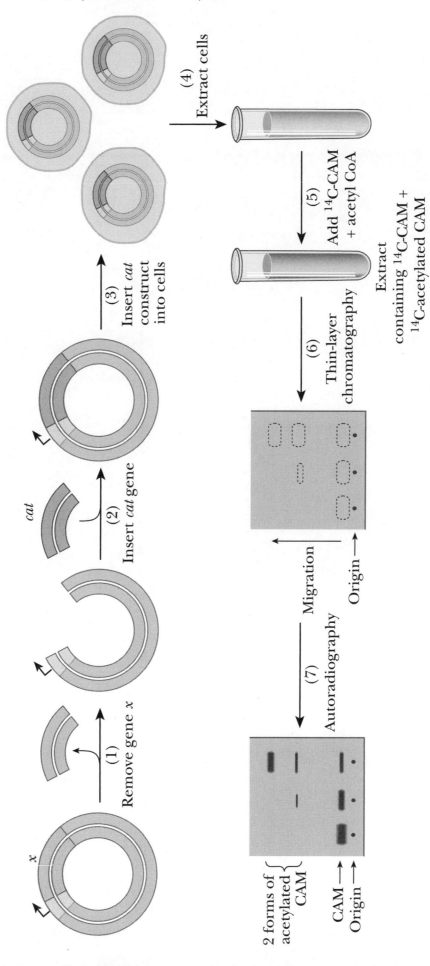

Figure 13.30 Using a reporter gene

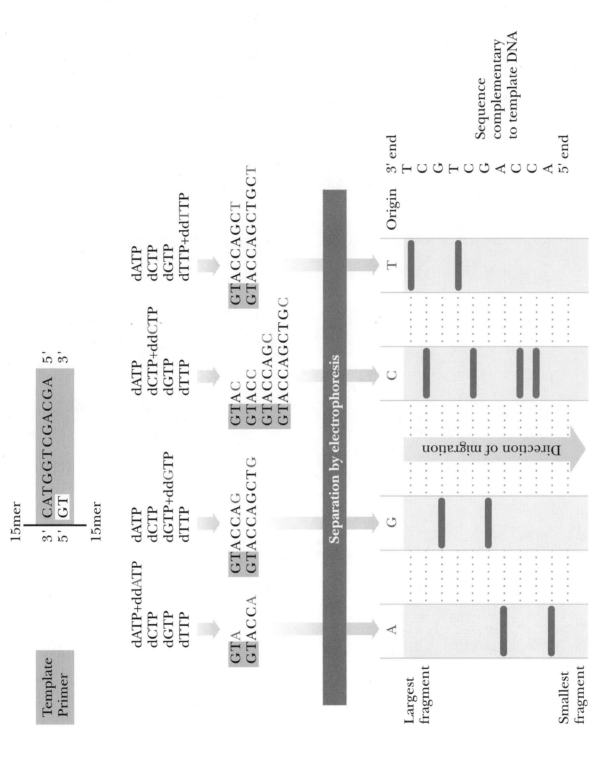

Figure 13.31 Sanger-Coulson method for DNA sequencing

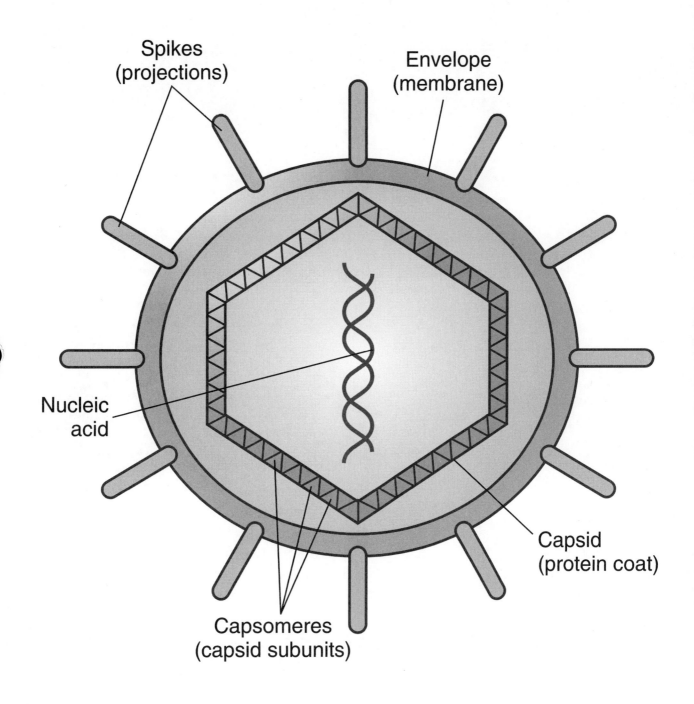

Spikes
(projections)

Envelope
(membrane)

Nucleic
acid

Capsid
(protein coat)

Capsomeres
(capsid subunits)

Figure 14.1 Architecture of a typical virus particle

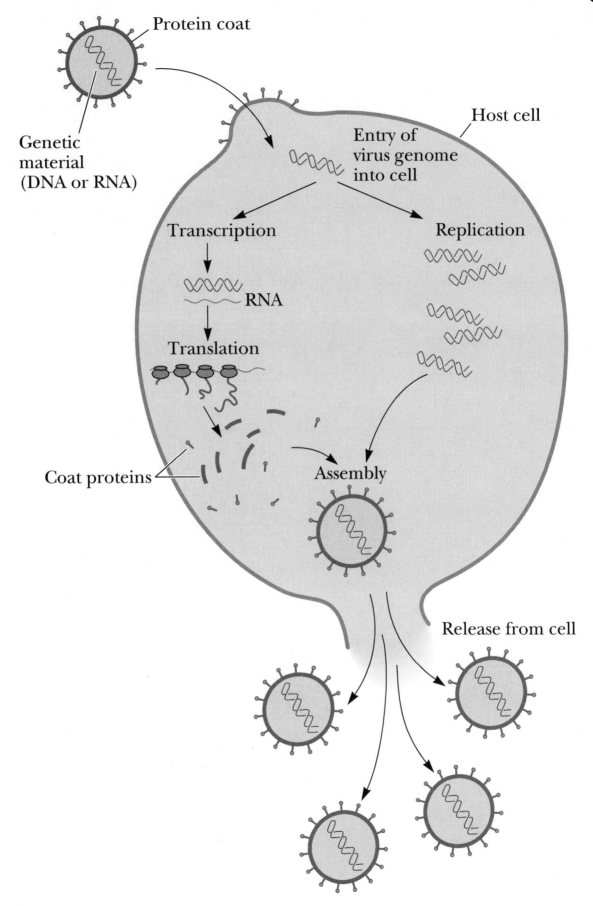

Figure 14.4 Virus life cycle

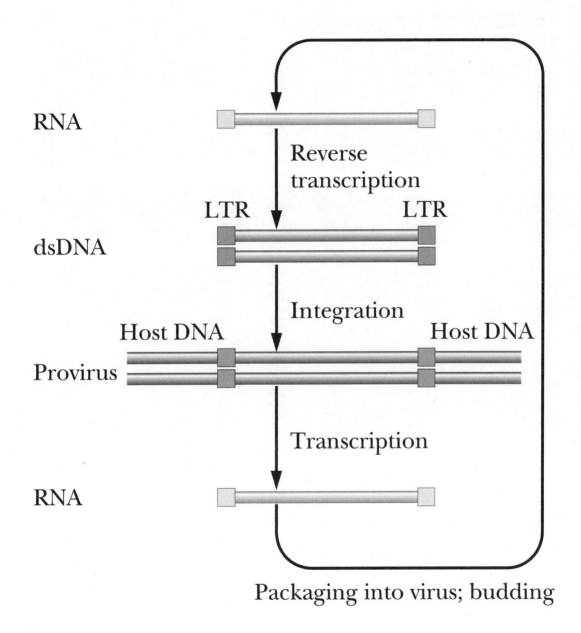

Figure 14.9 Life cycle of a retrovirus

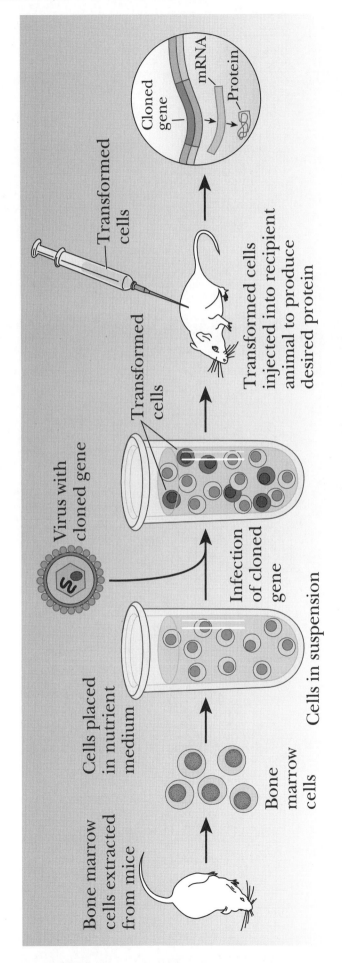

(Adapted from Dealing with Genes: The Language of Heredity, by Paul Berg and Maxine Singer, © 1992 by University Science Books.)

Figure 14.11 Gene therapy in bone marrow cells

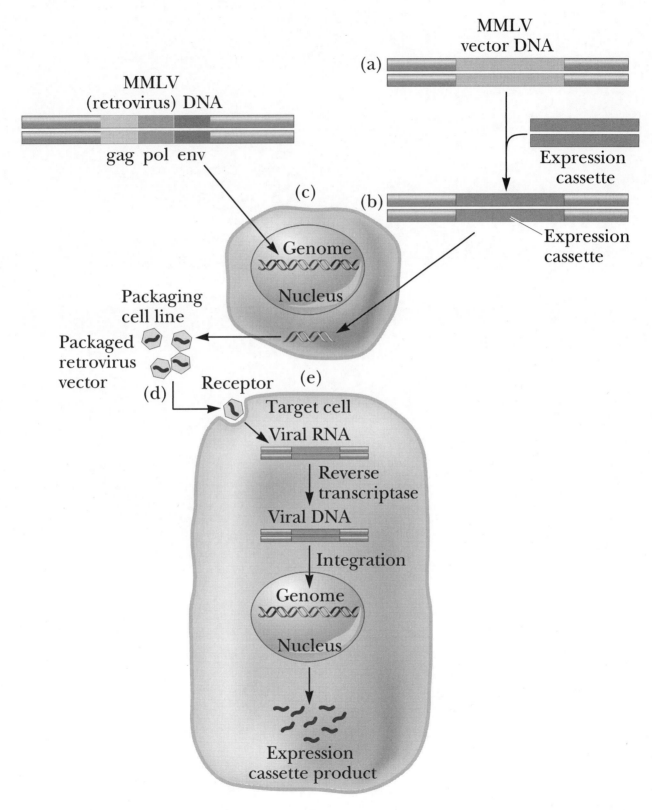

84

(*Adapted from Figure 1 in Crystal, R. G., 1995. Transfer of genes to humans: Early lessons and obstacles to success.* Science **270,** *404.*)

Figure 14.12 Human gene therapy via retroviruses

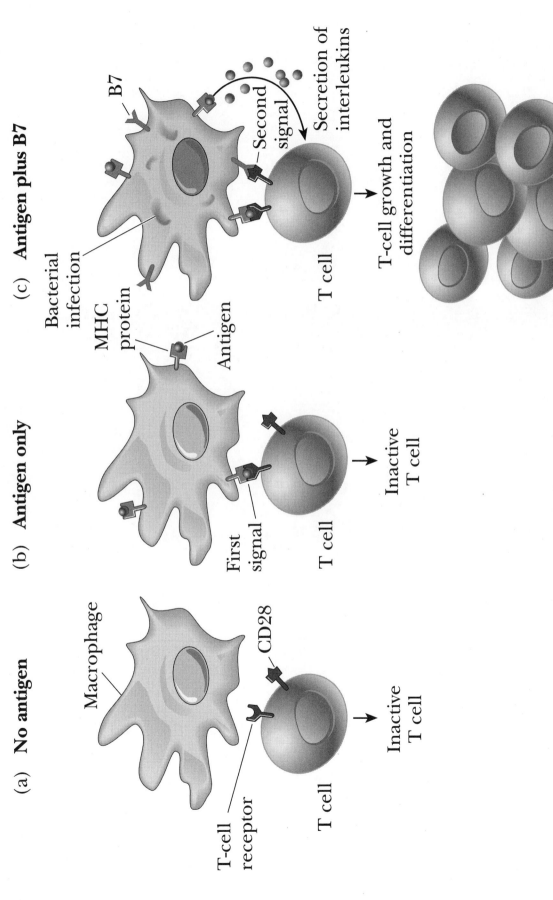

(a) **No antigen**

(b) **Antigen only**

(c) **Antigen plus B7**

Macrophage

T-cell receptor

CD28

T cell

Inactive T cell

First signal

MHC protein

Antigen

T cell

Inactive T cell

Bacterial infection

B7

Second signal

Secretion of interleukins

T-cell growth and differentiation

T cell

(Adapted from "How the Immune System Recognizes Invaders," by Charles A. Janeway, Jr.; illustration by Ian Warpole. Sci. Amer. 269 (3) (1993).)

Figure 14.18 A two-stage process leads to the growth and differentiation of T cells

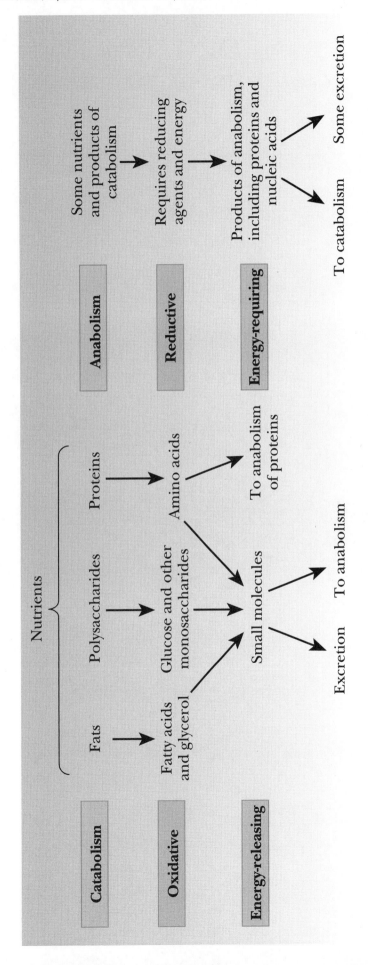

Figure 15.1 Catabolism and anabolism

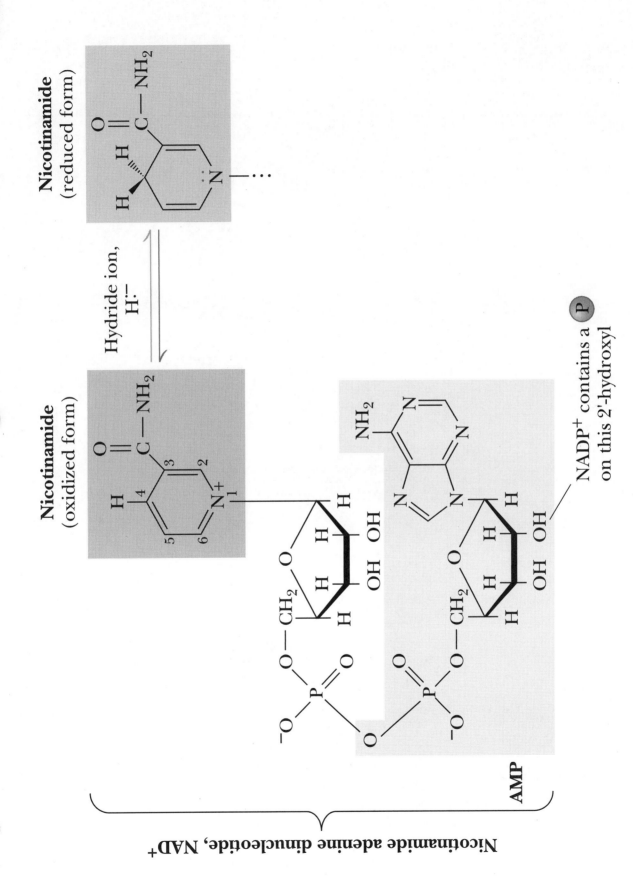

Figure 15.3 *NADH and NAD⁺*

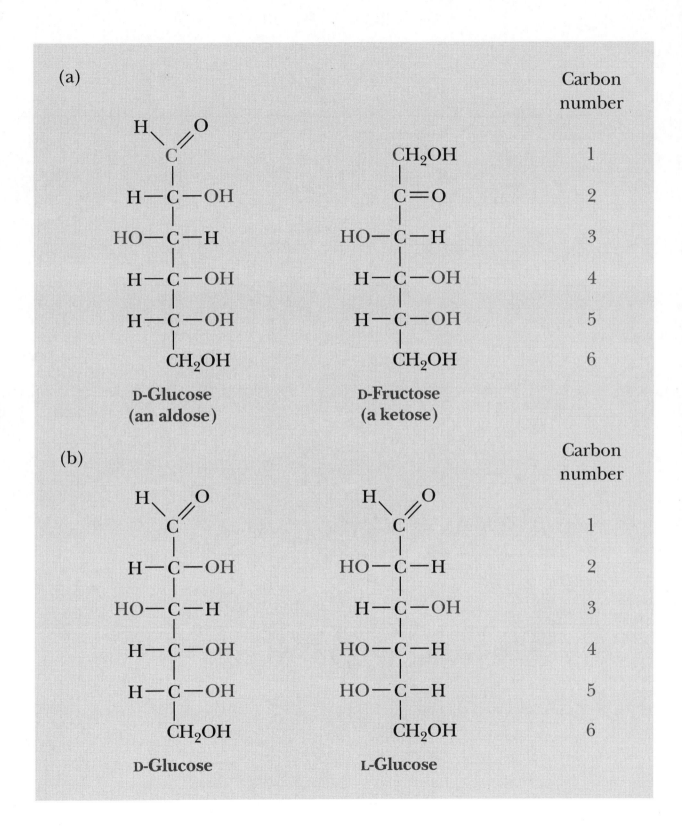

Figure 16.2 An aldose and a ketose

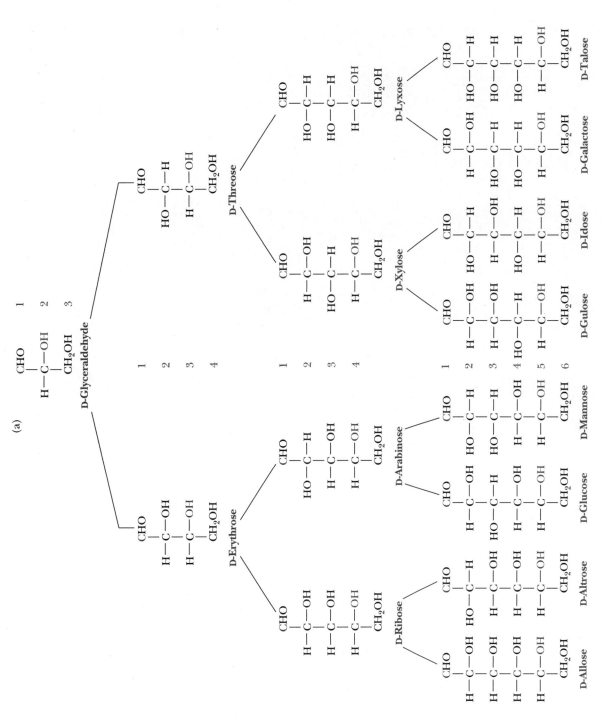

Figure 16.4a Stereochemical relationships among monosaccharides

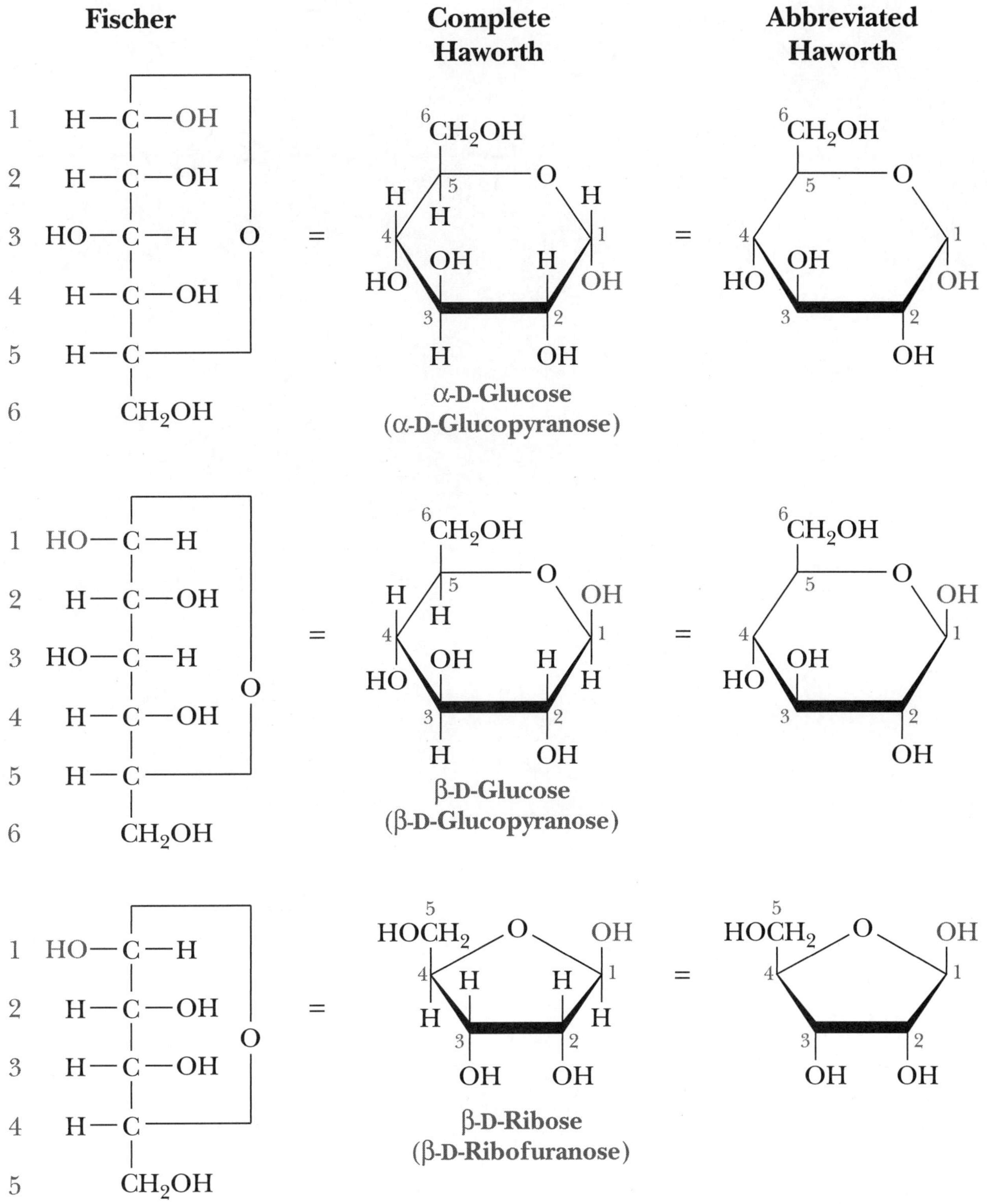

Figure 16.8 A comparison of the Fischer, complete Haworth, and abbreviated Haworth structures of α- and β-D-glucose and β-D-ribose

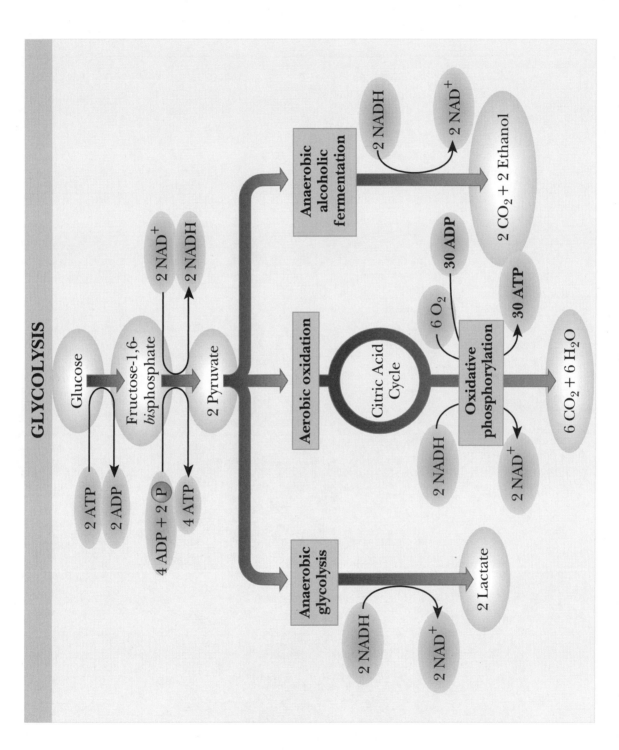

Figure 17.1 *One molecule of glucose is converted to two molecules of pyruvate*

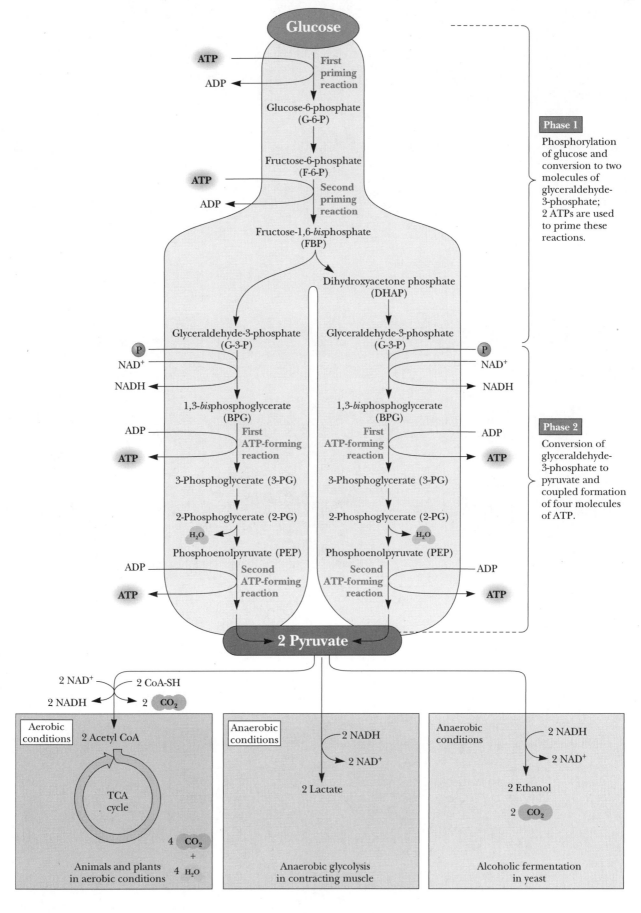

Figure 17.2 The pathway of glycolysis

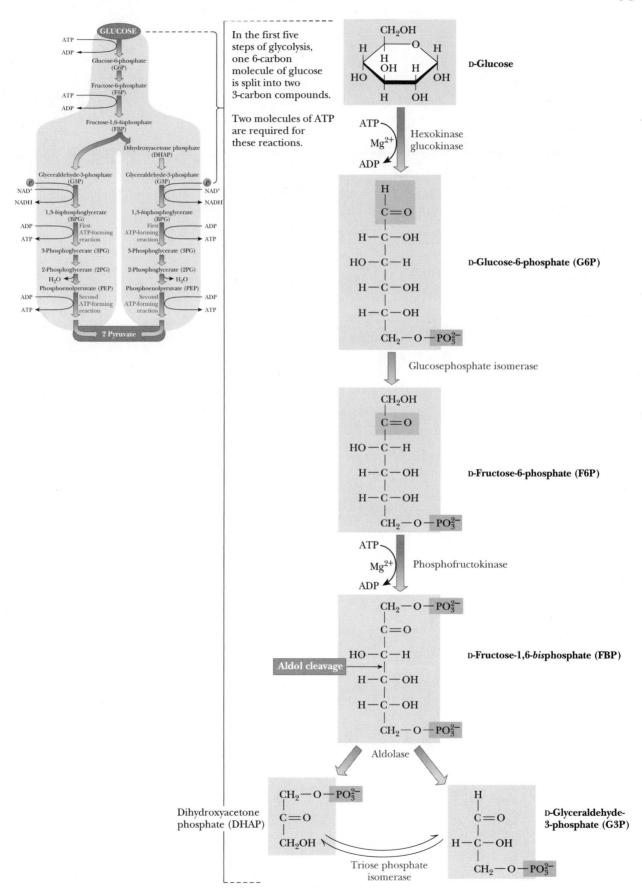

93

In the first five steps of glycolysis, one 6-carbon molecule of glucose is split into two 3-carbon compounds.

Two molecules of ATP are required for these reactions.

Figure 17.3 The first phase of glycolysis

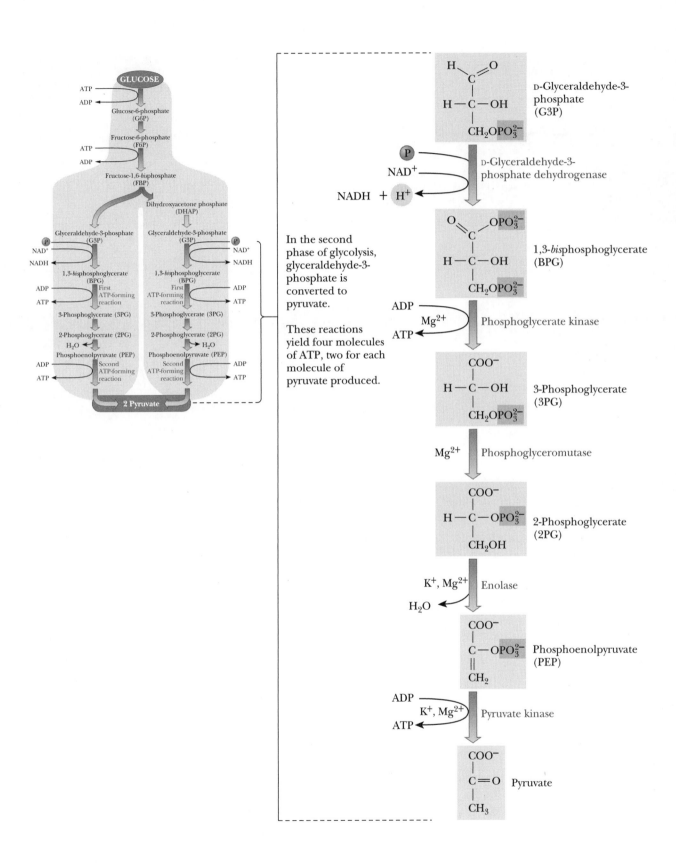

Figure 17.7 The second phase of glycolysis

Table 17.1

The Reactions of Glycolysis and Their Standard Free-Energy Changes

Step	Reaction	Enzyme	$\Delta G^{\circ\prime}$* kJ mol^{-1}	$\Delta G^{\circ\prime}$* kcal mol^{-1}	ΔG** kJ/mol
1	Glucose + ATP → Glucose-6-phosphate + ADP	Hexokinase/Glucokinase	−16.7	−4.0	−33.9
2	Glucose-6-phosphate → Fructose-6-phosphate	Glucose phosphate isomerase	+1.67	+0.4	−2.92
3	Fructose-6-phosphate + ATP → Fructose-1,6-bisphosphate + ADP	Phosphofructokinase	−14.2	−3.4	−18.8
4	Fructose-1,6-*bisphosphate* → Dihydroxyacetone phosphate + Glyceraldehyde-3-phosphate	Aldolase	+23.9	+5.7	−0.23
5	Dihydroxyacetone phosphate → Glyceraldehyde-3-phosphate	Triose phosphate isomerase	+7.56	+1.8	+2.41
6	2 (Glyceraldehyde-3-phosphate + NAD$^+$ + P$_i$ → 1,3-*bisphosphoglycerate* + NADH + H$^+$)	Glyceraldehyde-3-P dehydrogenase	2(+6.20)	2(+1.5)	2(−1.29)
7	2 (1,3-*bisphosphoglycerate* + ADP → 3-Phosphoglycerate + ATP)	Phosphoglycerate kinase	2(−18.8)	2(−4.5)	2(+0.1)
8	2 (3-Phosphoglycerate → 2-Phosphoglycerate)	Phosphoglyceromutase	2(+4.4)	2(+1.1)	2(+0.83)
9	2 (2-Phosphoglycerate → Phosphoenolpyruvate + H$_2$O)	Enolase	2(+1.8)	2(+0.4)	2(+1.1)
10	2 (Phosphoenolpyruvate + ADP → Pyruvate + ATP)	Pyruvate kinase	2(−31.4)	2(−7.5)	2(−23.0)
Overall	Glucose + 2 ADP + 2 P$_i$ + 2 NAD$^+$ → 2 Pyruvate → 2 ATP + NADH + H$^+$		−73.3	−17.5	−98.0
	2 Pyruvate + NADH + H$^+$ → Lactate + NAD$^+$	Lactate dehydrogenase	2(−25.1)	2(−6.0)	2(−14.8)
	Glucose + 2 ADP + 2 P$_i$ → 2 Lactate + 2 ATP		−123.5	−29.5	−127.6

*$\Delta G^{\circ\prime}$ values are assumed to be the same at 25°C and 37°C and are calculated for standard-state conditions (1 M concentration of reactants and products, pH 7.0).

**ΔG values are calculated at 310 K (37°C) using steady-state concentrations of these metabolites found in erythrocytes.

Table 17.1 *The reactions of glycolysis and their standard free-energy changes*

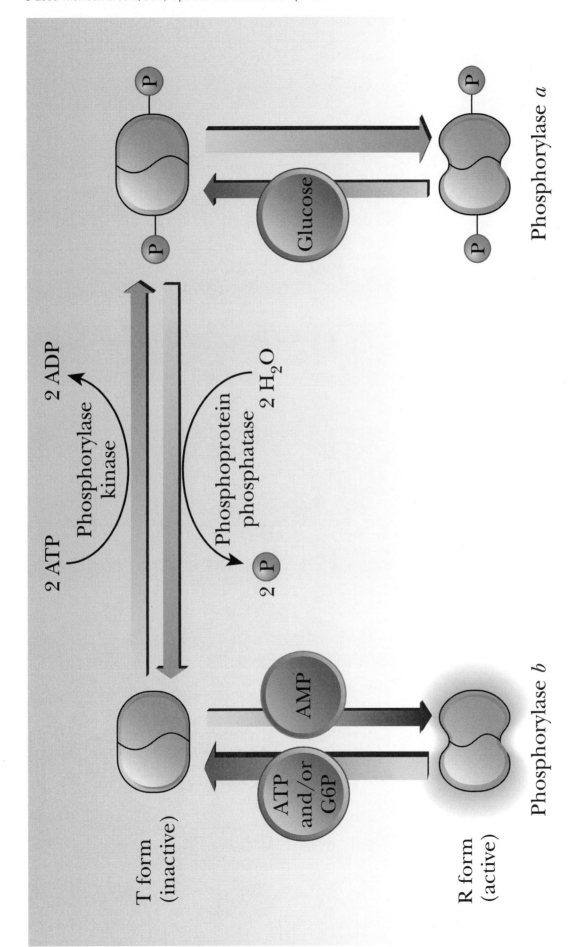

Figure 18.5 Glycogen phosphorylase activity is subject to allosteric control and covalent modification

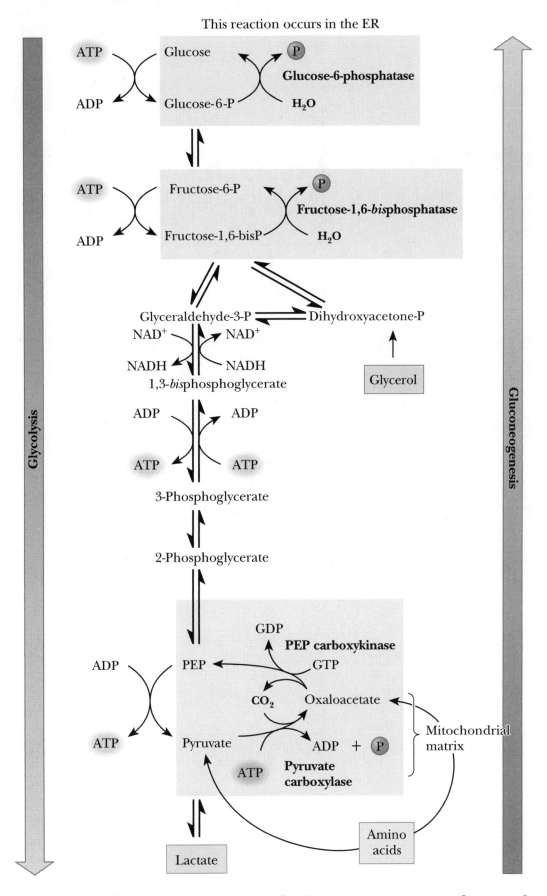

Figure 18.6 The pathways of gluconeogenesis and glycolysis

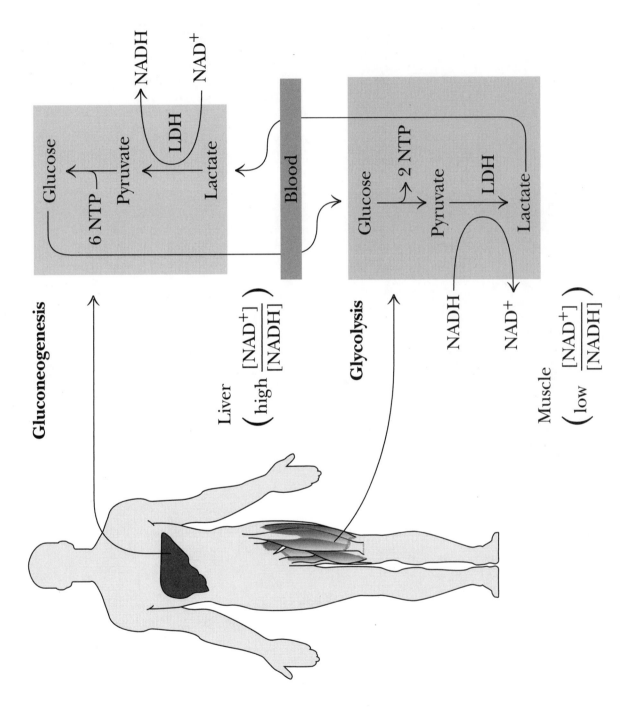

Figure 18.12 *The Cori cycle*

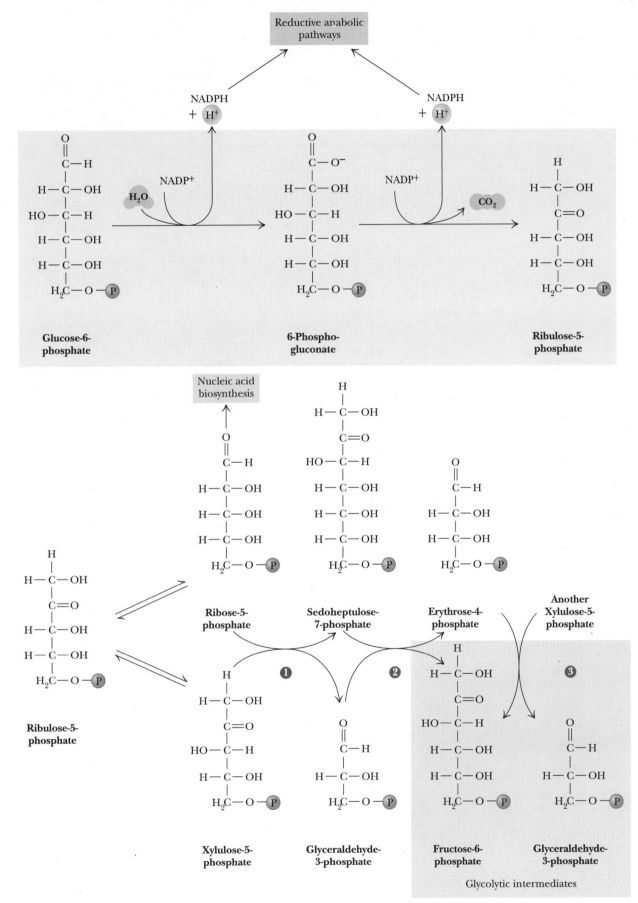

Figure 18.15 **The pentose phosphate pathway**

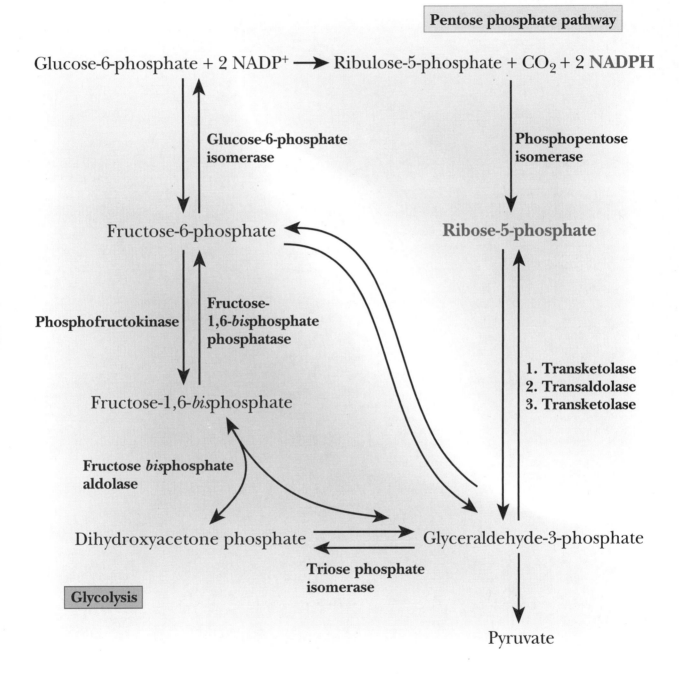

Figure 18.16 Relationships between the pentose phosphate pathway and glycolysis

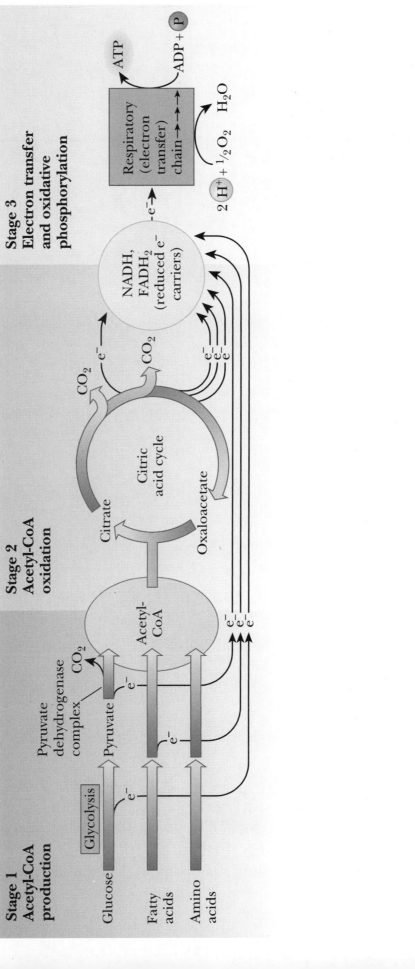

Figure 19.1 The citric acid cycle and catabolism

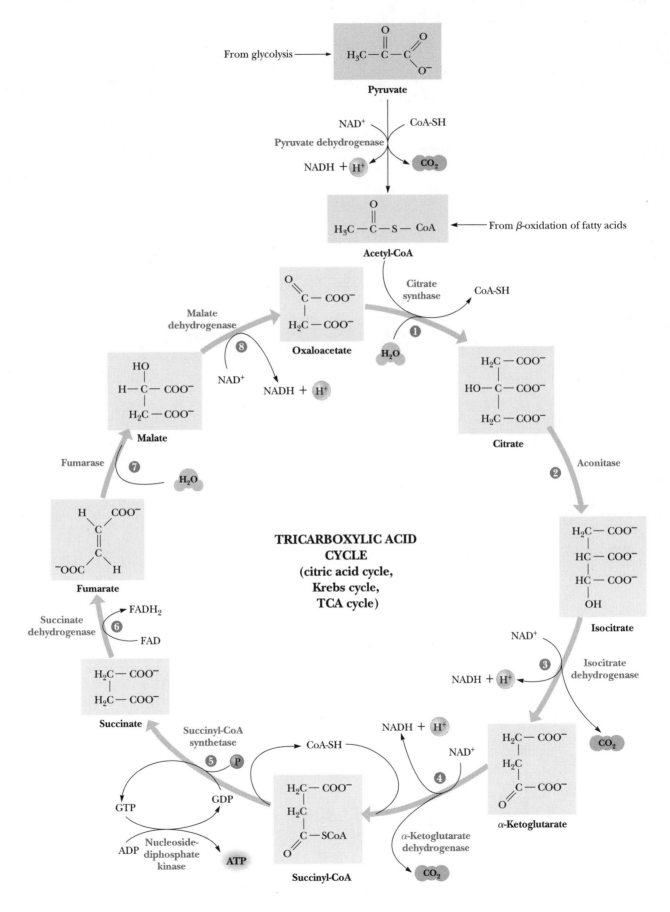

Figure 19.3 An overview of the citric acid cycle

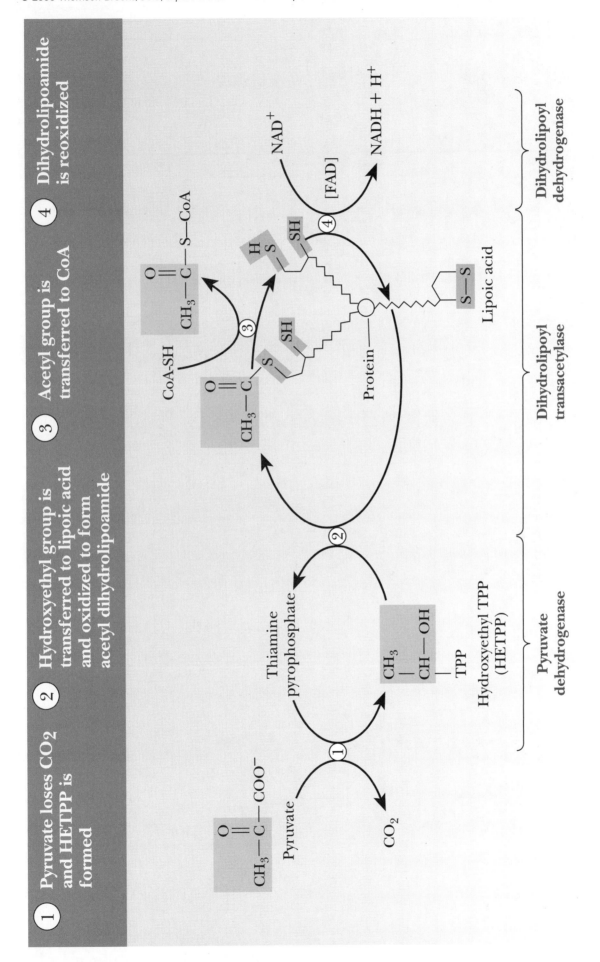

Figure 19.4 The mechanism of the pyruvate dehydrogenase reaction

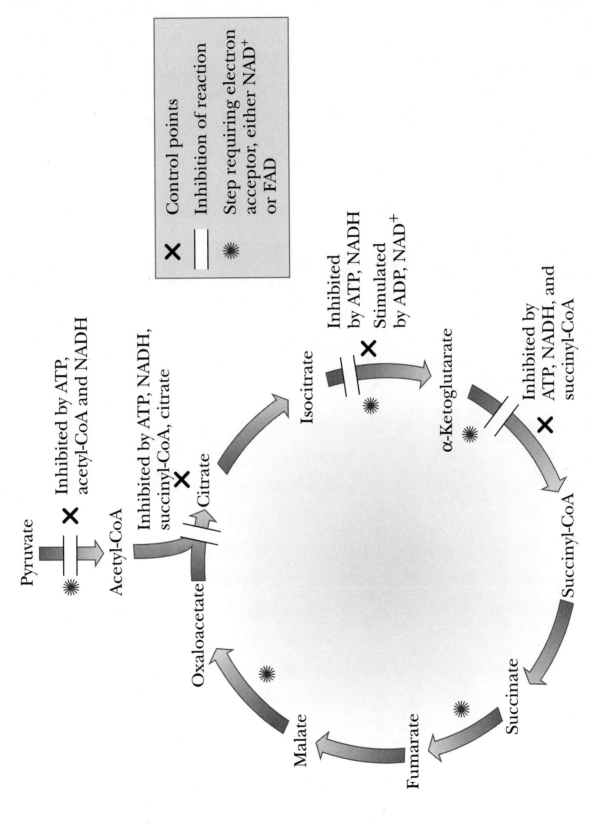

Figure 19.8 Control points in the conversion of pyruvate to acetyl-CoA and in the citric acid cycle

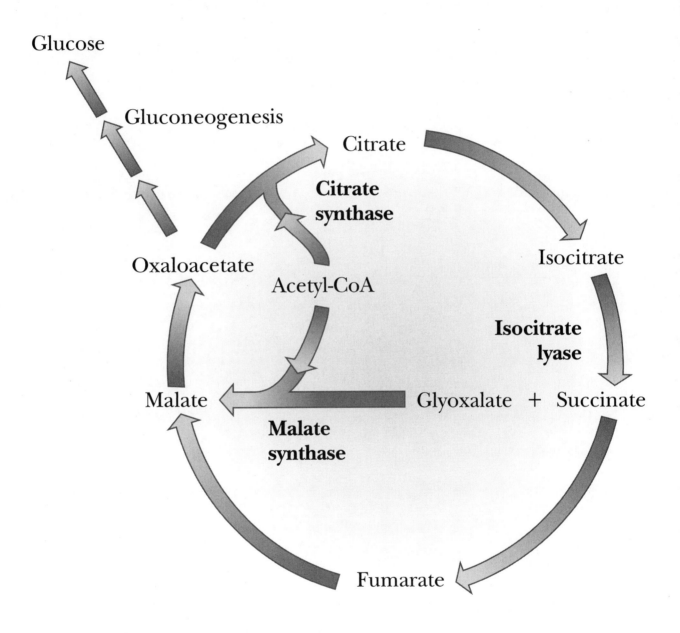

Figure 19.9 The glyoxylate cycle

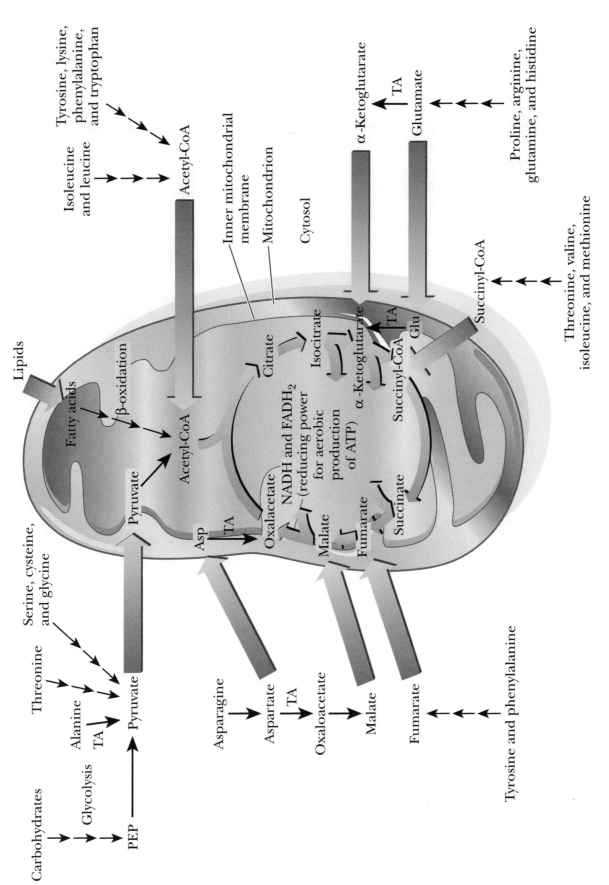

Figure 19.10 A summary of catabolism

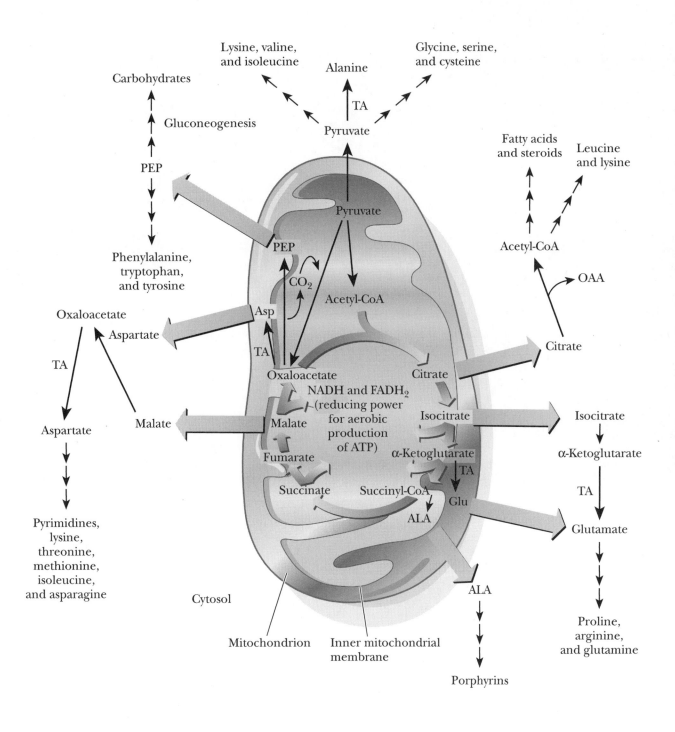

Figure 19.15 A summary of anabolism

108

Table 19.2

The Energetics of Conversion of Pyruvate to CO$_2$

Step	Reaction	$\Delta G^{\circ\prime}$ (kJ mol^{-1})	$\Delta G^{\circ\prime}$ (kcal mol^{-1})
	Pyruvate + CoA-SH + NAD$^+$ → Acetyl-CoA + NADH + CO$_2$	−33.4	−8.0
1	Acetyl-CoA + Oxaloacetate + H$_2$O → Citrate + CoA-SH + H$^+$	−32.2	−7.7
2	Citrate → Isocitrate	+6.3	+1.5
3	Isocitrate + NAD$^+$ → α-Ketoglutarate + NADH + CO$_2$ + H$^+$	−7.1	−1.7
4	α-Ketoglutarate + NAD$^+$ + CoA-SH → Succinyl-CoA + NADH + CO$_2$ + H	−33.4	−8.0
5	Succinyl-CoA + GDP + P$_i$ → Succinate + GTP + CoA-SH	−3.3	−0.8
6	Succinate + FAD → Fumarate + FADH$_2$	~0	~0
7	Fumarate + H$_2$O → L-Malate	−3.8	−0.9
8	L-Malate + NAD$^+$ → Oxaloacetate + NADH + H$^+$	+29.2	+7.0
Overall:	Pyruvate + 4 NAD$^+$ + FAD + GDP + P$_i$ + 2 H$_2$O → CO$_2$ + 4 NADH + FADH$_2$ + GTP + 4 H$^+$	−77.7	−18.6

Table 19.2 *The energetics of conversion of pyruvate to carbon dioxide*

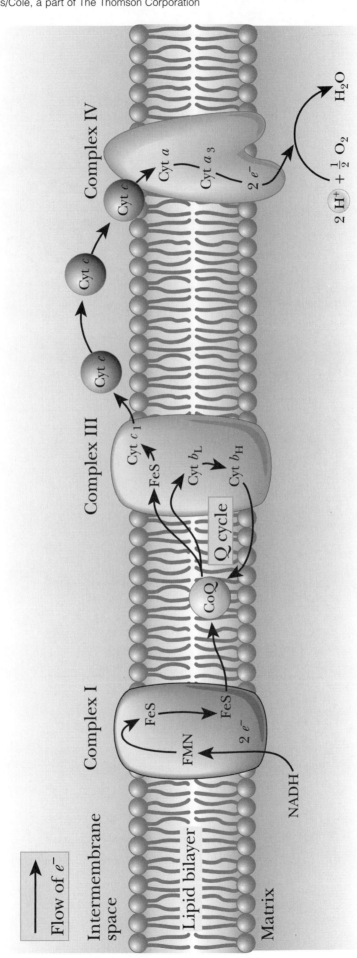

109

Figure 20.7 Respiratory complexes in the inner mitochondrial membrane

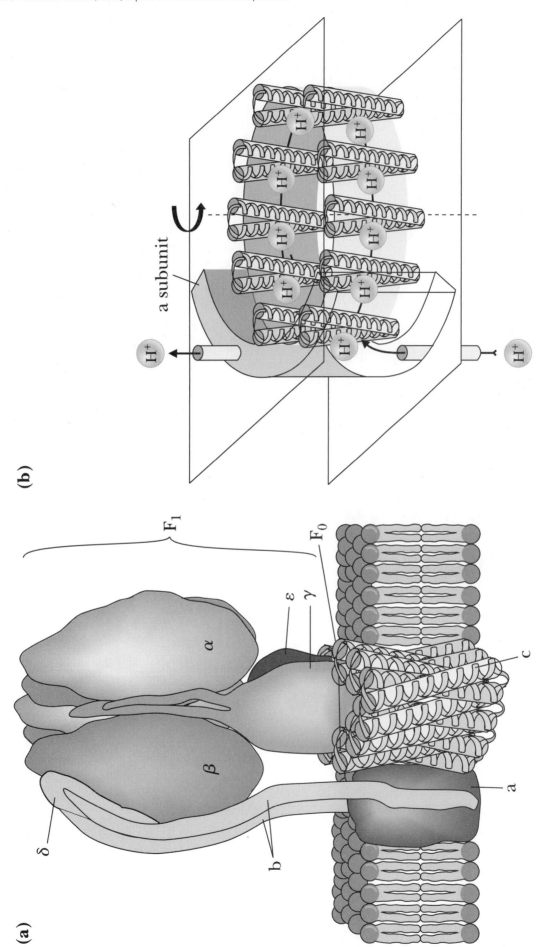

(b)

(a)

Figure 20.11 A model for the components of ATP synthase

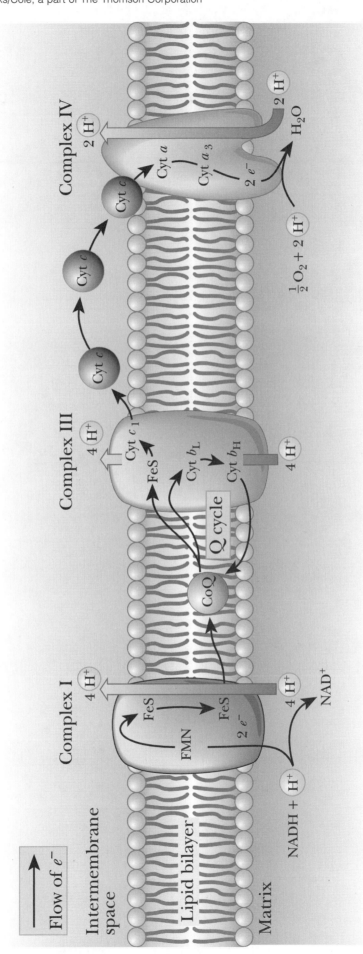

111

Figure 20.13 The creation of a proton gradient in chemiosmotic coupling

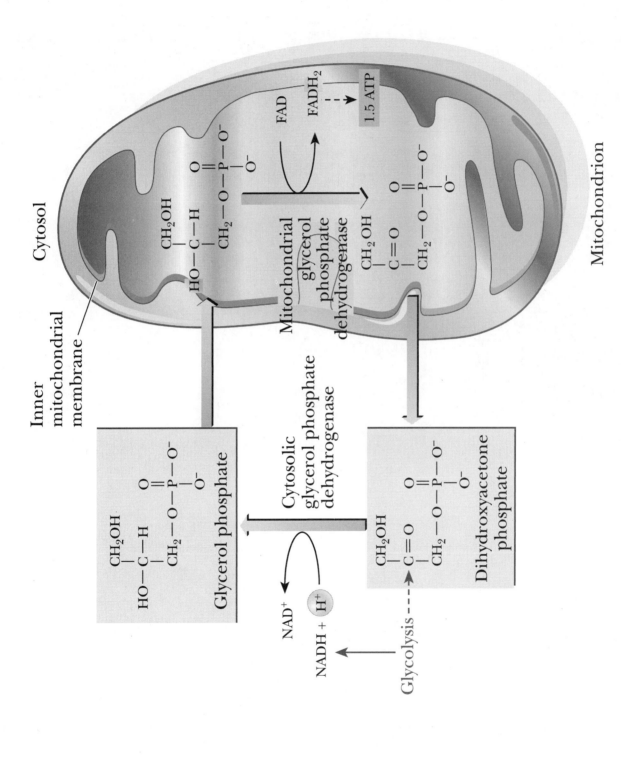

Figure 20.21 The glycerol-phosphate shuttle

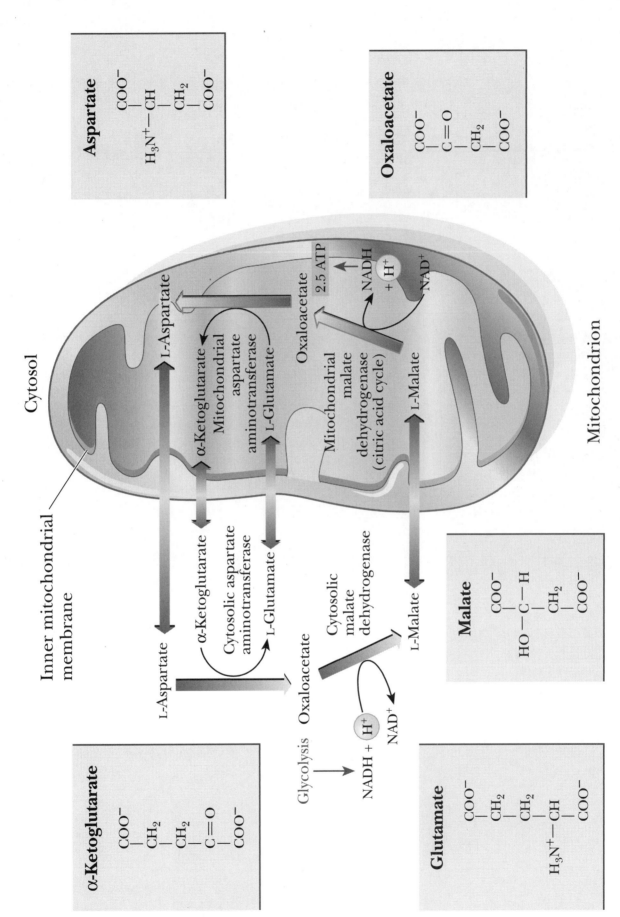

Figure 20.22 The malate-aspartate shuttle

114

Table 20.1

Standard Reduction Potentials for Several Biological Reduction Half-Reactions

Reduction Half-Reaction	$E^{\circ\prime}$ (V)
$\frac{1}{2} O_2 + 2 H^+ + 2 e^- \rightarrow H_2O$	0.816
$Fe^{3+} + e^- \rightarrow Fe^{2+}$	0.771
Cytochrome $a_3(Fe^{3+}) + e^- \rightarrow$ Cytochrome $a_3(Fe^{2+})$	0.350
Cytochrome $a(Fe^{3+}) + e^- \rightarrow$ Cytochrome $a(Fe^{2+})$	0.290
Cytochrome $c(Fe^{3+}) + e^- \rightarrow$ Cytochrome $c\ (Fe^{2+})$	0.254
Cytochrome $c_1(Fe^{3+}) + e^- \rightarrow$ Cytochrome $c_1(Fe^{2+})$	0.220
$CoQH^{\cdot} + H^+ + e^- \rightarrow CoQH_2$ (coenzyme Q)	0.190
$CoQ + 2 H^+ + 2 e^- \rightarrow CoQH_2$	0.060
Cytochrome $b_H(Fe^{3+}) + e^- \rightarrow$ Cytochrome $b_H(Fe^{2+})$	0.050
Fumarate $+ 2 H^+ + 2 e^- \rightarrow$ Succinate	0.031
$CoQ + H^+ + e^- \rightarrow CoQH^{\cdot}$	0.030
$[FAD] + 2 H^+ + 2 e^- \rightarrow [FADH_2]$	0.003–0.091*
Cytochrome $b_L(Fe^{3+}) + e^- \rightarrow$ Cytochrome $b_L(Fe^{2+})$	−0.100
Oxaloacetate $+ 2 H^+ + e^- \rightarrow$ Malate	−0.166
Pyruvate $+ 2 H^+ + 2 e^- \rightarrow$ Lactate	−0.185
Acetaldehyde $+ 2 H^+ + 2 e^- \rightarrow$ Ethanol	−0.197
$FMN + 2 H^+ + 2 e^- \rightarrow FMNH_2$	−0.219
$FAD + 2 H^+ + 2 e^- \rightarrow FADH_2$	−0.219
1,3-*bis*phosphoglycerate $+ 2 H^+ + 2 e^- \rightarrow$	
Glyceraldehyde-3-phosphate $+ P_i$	−0.290
$NAD^+ + 2 H^+ + 2 e^- \rightarrow NADH + H^+$	−0.320
$NADP^+ + 2 H^+ + 2 e^- \rightarrow NADPH + H^+$	−0.320
α-Ketoglutarate $+ CO_2 + 2 H^+ + 2 e^- \rightarrow$ Isocitrate	−0.380
Succinate $+ CO_2 + 2 H^+ + 2 e^- \rightarrow \alpha$-Ketoglutarate $+ H_2O$	−0.670

* Typical values for reduction of bound FAD in flavoproteins such as succinate dehydrogenase.

Note that we have shown a number of components of the electron transport chain individually. We are going to see them again as parts of complexes. We have also included values for a number of reactions we saw in earlier chapters.

Table 20.1 Standard reduction potentials for biological half-reactions

115

Table 20.2

The Energetics of Electron Transport Reactions

Reaction	$\Delta G^{\circ\prime}$	
	kJ $(\text{mol NADH})^{-1}$	$kcal$ $(\text{mol NADH})^{-1}$
$NADH + H^+ + E{-}FMN \rightarrow NAD^+ + E{-}FMNH_2$	-38.6	-9.2
$E{-}FMNH_2 + CoQ \rightarrow E{-}FMN + CoQH_2$	-42.5	-10.2
$CoQH_2 + 2\ \text{Cyt}\ b[Fe(III)] \rightarrow CoQ + 2\ H^+ + 2\ \text{Cyt}\ b[Fe(II)]$	$+11.6$	$+2.8$
$2\ \text{Cyt}\ b[Fe(II)] + 2\ \text{Cyt}\ c_1[Fe(III)] \rightarrow 2\ \text{Cyt}\ c_1[Fe(II)] + 2\ \text{Cyt}\ b[Fe(III)]$	-34.7	-8.3
$2\ \text{Cyt}\ c_1[Fe(II)] + 2\ \text{Cyt}\ c[Fe(III)] \rightarrow 2\ \text{Cyt}\ c[Fe(II)] + 2\ \text{Cyt}\ c_1[Fe(III)]$	-5.8	-1.4
$2\ \text{Cyt}\ c[Fe(II)] + 2\ \text{Cyt}\ (aa_3)\ [Fe(III)] \rightarrow 2\ \text{Cyt}\ (aa_3)\ [Fe(II)] + 2\ \text{Cyt}\ c[Fe(III)]$	-7.7	-1.8
$2\ \text{Cyt}\ (aa_3)\ [Fe(II)] + \frac{1}{2}O_2 + 2\ H^+ \rightarrow 2\ \text{Cyt}\ (aa_3)\ [Fe(III)] + H_2O$	-102.3	-24.5
Overall reaction: $NADH + H^+ + \frac{1}{2}O_2 \rightarrow NAD^+ + H_2O$	-220	-52.6

Table 20.2 Energetics of electron transport reactions

116

Table 20.3

Yield of ATP from Glucose Oxidation

Pathway	ATP Yield per Glucose		NADH	FADH$_2$
	Glycerol–Phosphate Shuttle	Malate–Aspartate Shuttle		
Glycolysis: glucose to pyruvate (cytosol)				
Phosphorylation of glucose	−1	−1		
Phosphorylation of fructose-6-phosphate	−1	−1		
Dephosphorylation of 2 molecules of 1,3-BPG	+2	+2		
Dephosphorylation of 2 molecules of PEP	+2	+2		
Oxidation of 2 molecules of glyceraldehyde-3-phosphate yields 2 NADH			+2	
Pyruvate conversion to acetyl-CoA (mitochondria)				
2 NADH produced			+2	
Citric acid cycle (mitochondria)				
2 molecules of GTP from 2 molecules of succinyl-CoA	+2	+2		
Oxidation of 2 molecules each of isocitrate, α-ketoglutarate, and malate yields 6 NADH			+6	
Oxidation of 2 molecules of succinate yields 2 FADH$_2$				+2
Oxidative phosphorylation (mitochondria)				
2 NADH from glycolysis yield 1.5 ATP each if NADH is oxidized by glycerol–phosphate shuttle; 2.5 ATP by malate–aspartate shuttle	+3	+5	−2	
Oxidative decarboxylation of 2 pyruvate to 2 acetyl-CoA: 2 NADH produce 2.5 ATP each	+5	+5	−2	
2 FADH$_2$ from each citric acid cycle produce 1.5 ATP each	+3	+3		−2
6 NADH from citric acid cycle produce 2.5 ATP each	+15	+15	−6	
Net Yield	+30	+32	0	0

(*Note:* These P/O ratios of 2.5 and 1.5 for mitochondrial oxidation of NADH and FADH$_2$ are "consensus values." Since they may not reflect actual values, and since these ratios may change depending on metabolic conditions, these estimates of ATP yield from glucose oxidation are approximate.)

Table 20.3 Yield of ATP from glucose oxidation

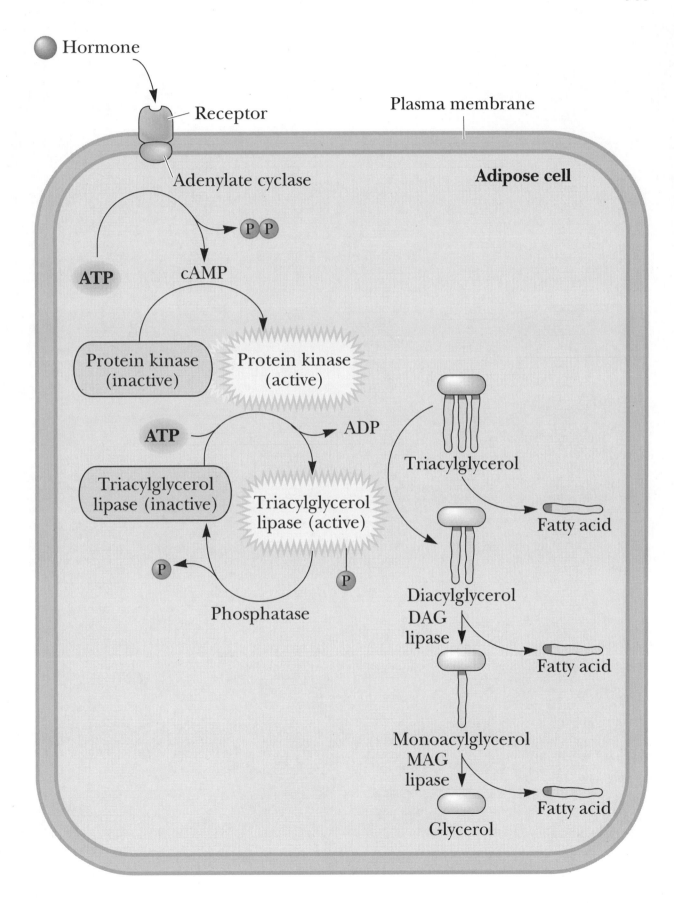

Figure 21.3 Liberation of fatty acids from triacylglycerols

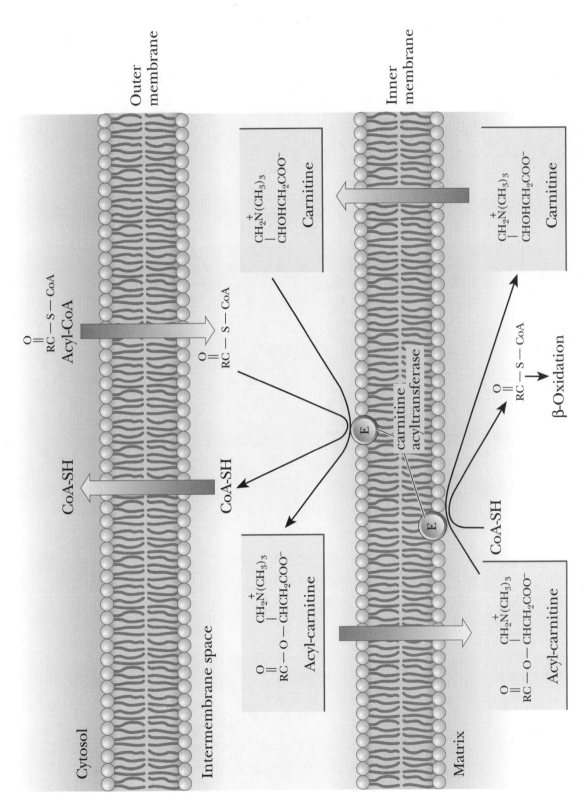

Figure 21.5 The role of carnitine in the transfer of acyl groups to the mitochondrial matrix

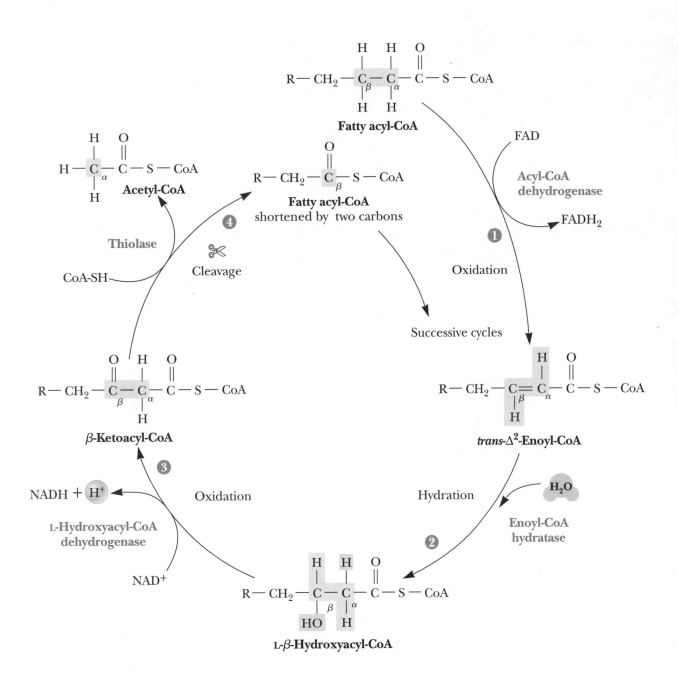

Figure 21.6 **The β-oxidation of saturated fatty acids**

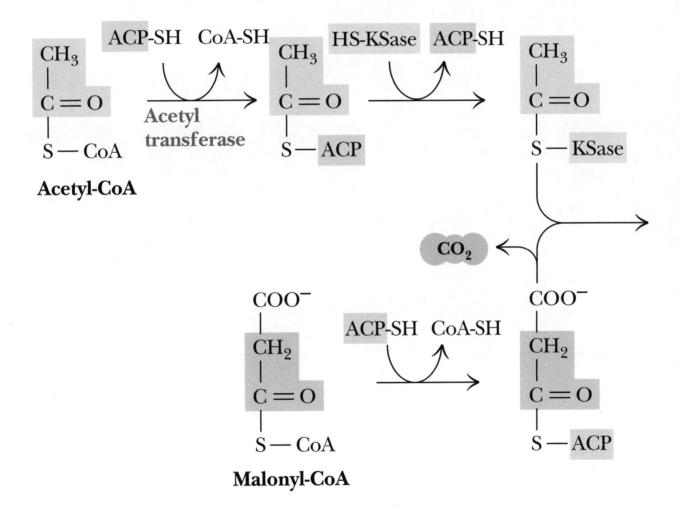

Figure 21.15 (left) The pathway of palmitate synthesis

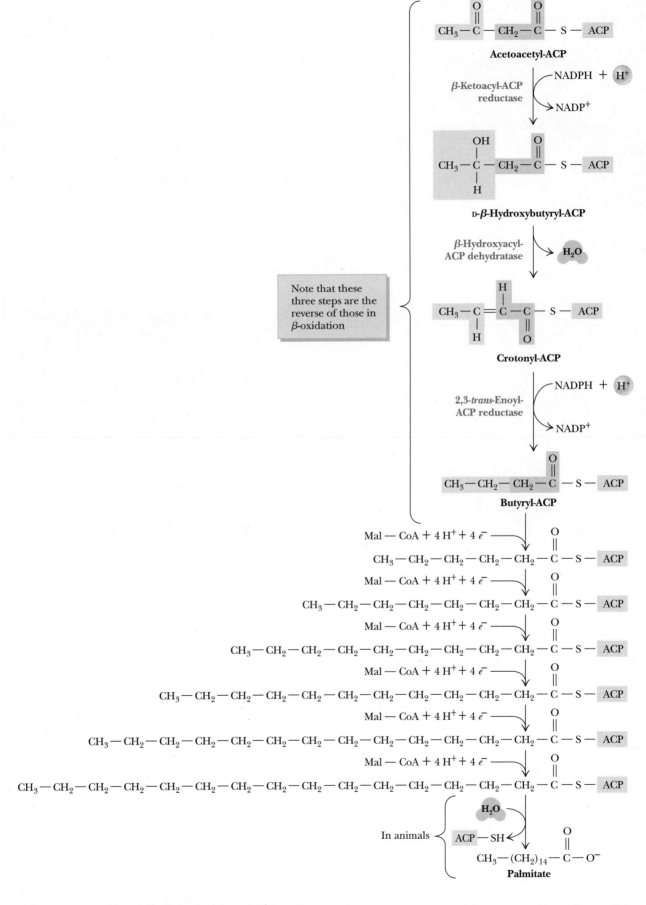

Figure 21.15 (right) The pathway of palmitate synthesis

Table 21.1

The Balance Sheet for Oxidation of One Molecule of Stearic Acid

Reaction	NADH Molecules	FADH$_2$ Molecules	ATP Molecules
1. Stearic acid → Stearyl-CoA (activation step)			−2
2. Stearyl-CoA → 9 acetyl-CoA (8 cycles of β-oxidation)	+8	+8	
3. 9 Acetyl-CoA → 18 CO$_2$ (citric acid cycle); GDP → GTP (9 molecules)	+27	+9	+9
4. Reoxidation of NADH from β-oxidation cycle	−8		+20
5. Reoxidation of NADH from citric acid cycle	−27		+67.5
6. Reoxidation of FADH$_2$ from β-oxidation cycle		−8	+12
7. Reoxidation of FADH$_2$ from citric acid cycle		−9	+13.5
	0	0	+120

Note that there is no net change in the number of molecules of NADH or FADH$_2$.

Table 21.1 The balance sheet for oxidation of one molecule of stearic acid

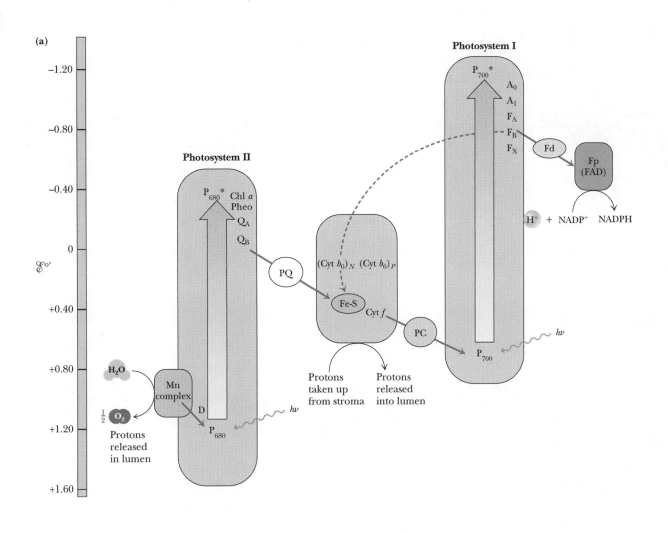

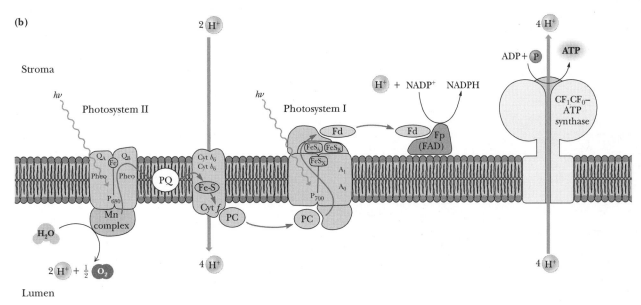

Figure 22.6 Electron flow in Photosystems I and II

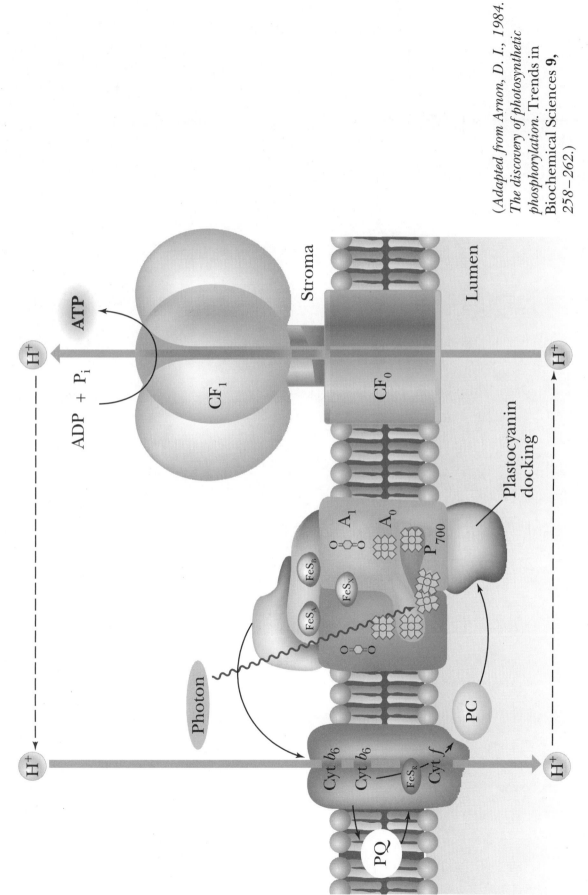

124

(Adapted from Arnon, D. I., 1984. The discovery of photosynthetic phosphorylation. Trends in Biochemical Sciences **9**, *258–262.)*

Figure 22.9 Cyclic electron flow coupled to photophosphorylation in Photosystem I

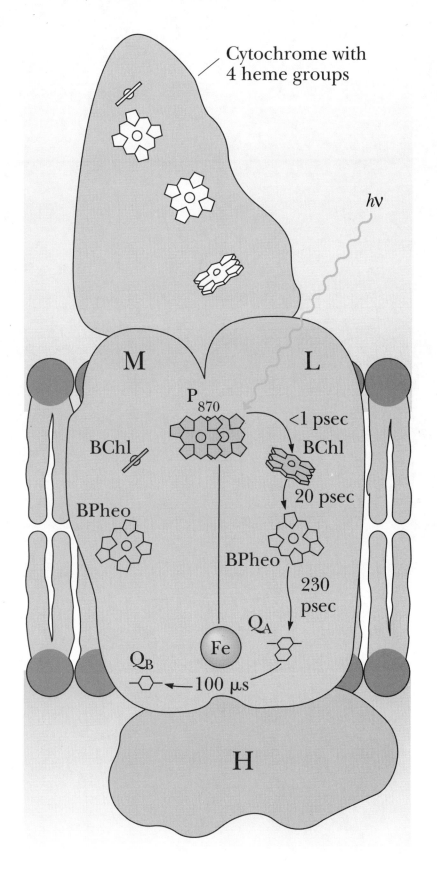

Figure 22.10 (top) *Molecular events that take place at the photosynthetic reaction center of* Rhodopseudomonas

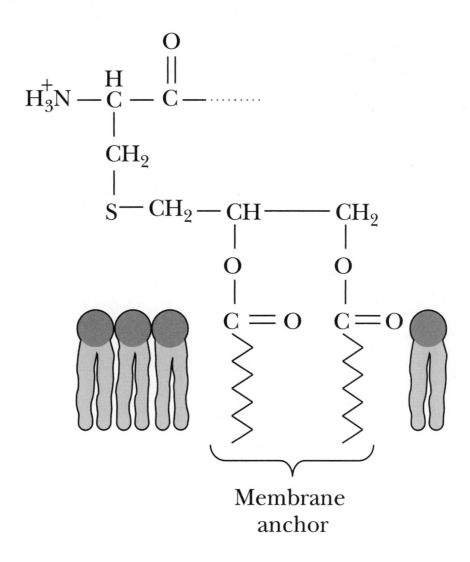

Figure 22.10 (bottom) Molecular events that take place at the photosynthetic reaction center of **Rhodopseudomonas**

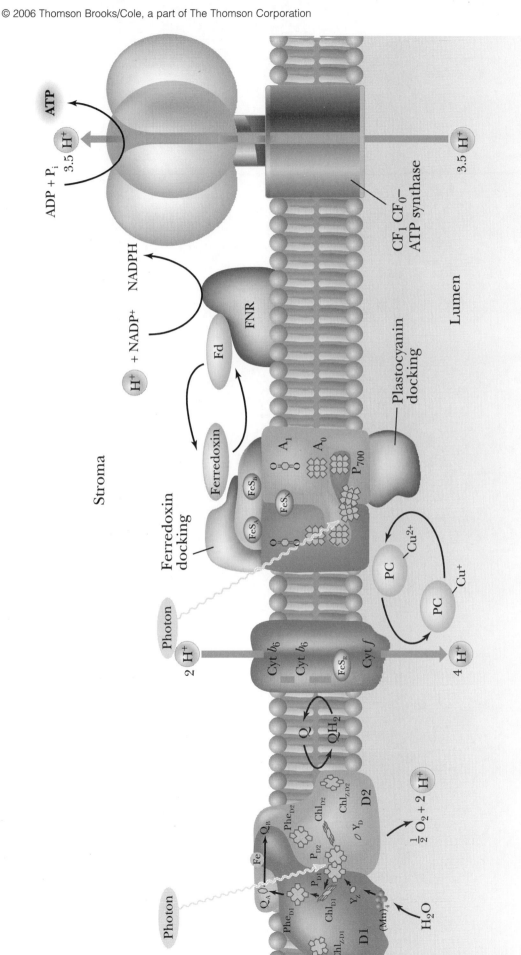

Figure 22.12 *The mechanism of photophosphorylation*

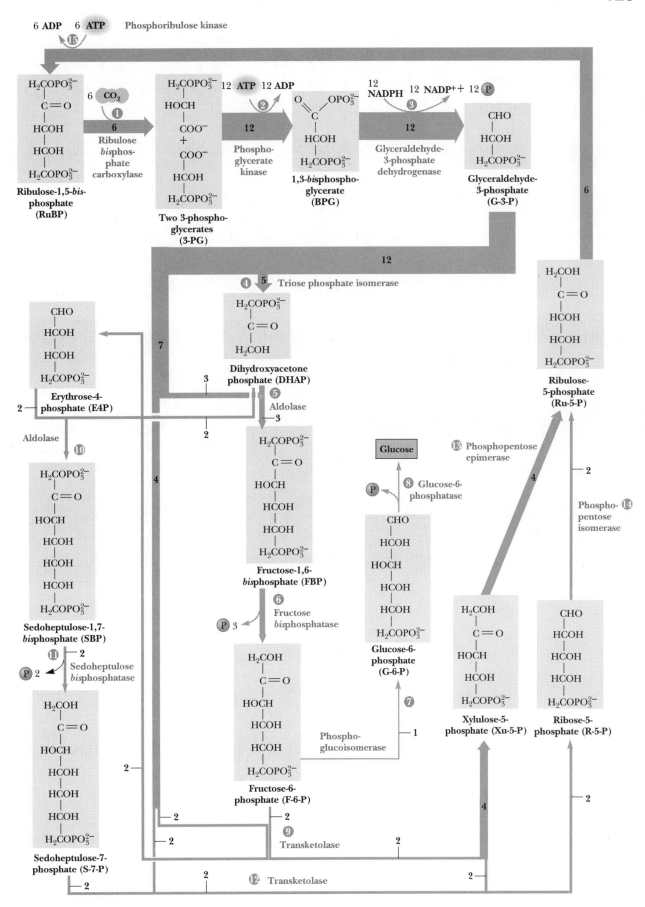

Figure 22.14 The Calvin cycle

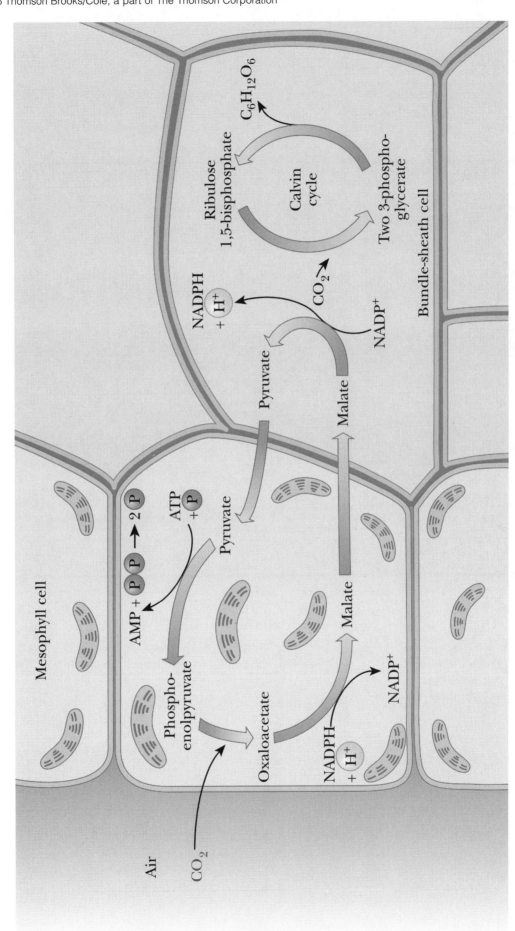

Figure 22.17 The C₄ pathway

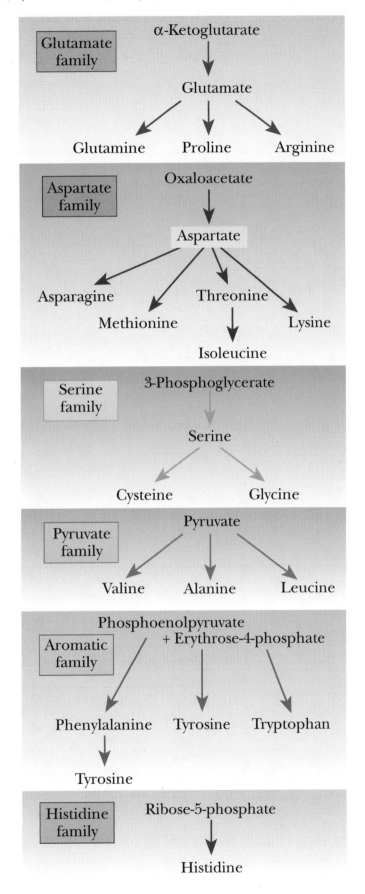

Figure 23.6 Families of amino acids based on biosynthetic pathways

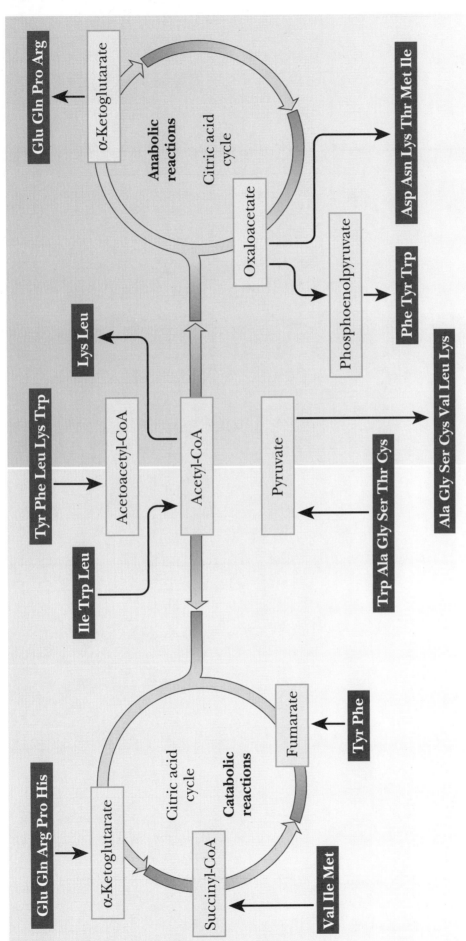

Anabolic formation of amino acids uses citric acid cycle intermediates as precursors

Catabolic breakdown of amino acids produces citric acid cycle intermediates

Figure 23.7 The relationship between amino acid metabolism and the citric acid cycle

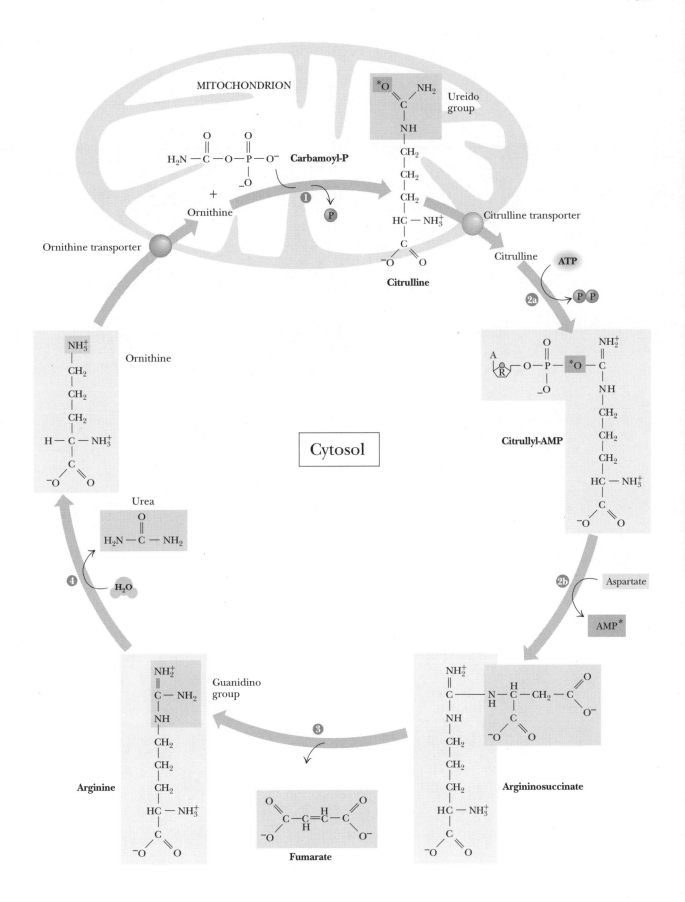

Figure 23.18 The urea cycle

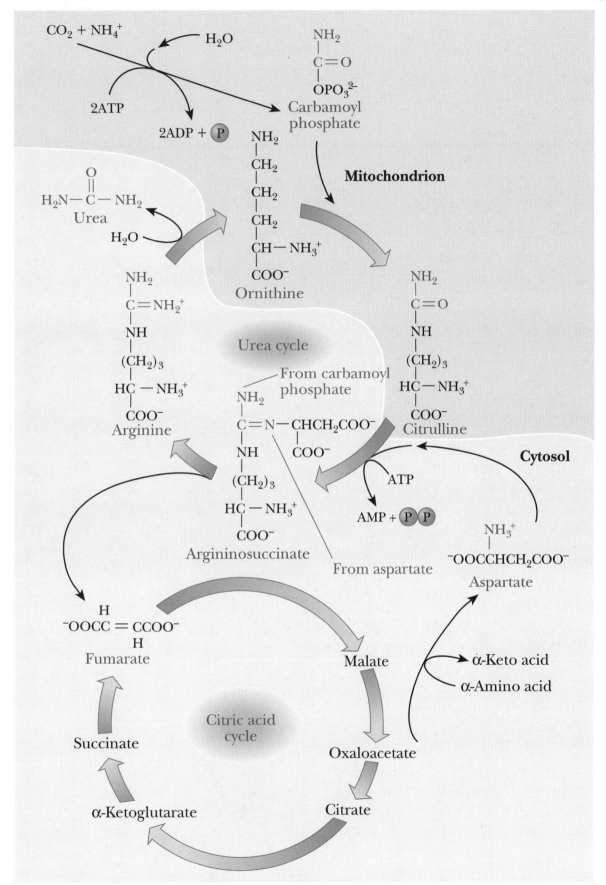

Figure 23.19 The urea cycle and some of its links to the citric acid cycle

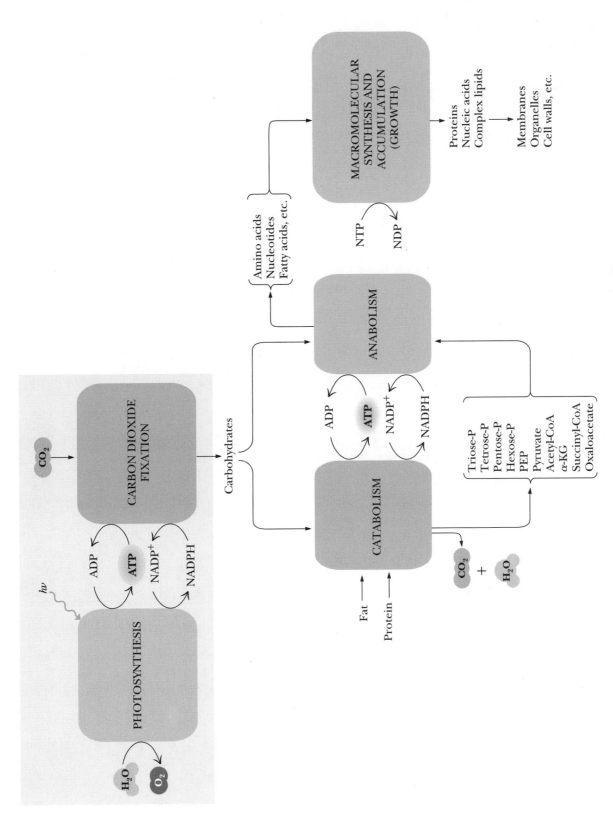

Figure 24.1 Block diagram of intermediary metabolism

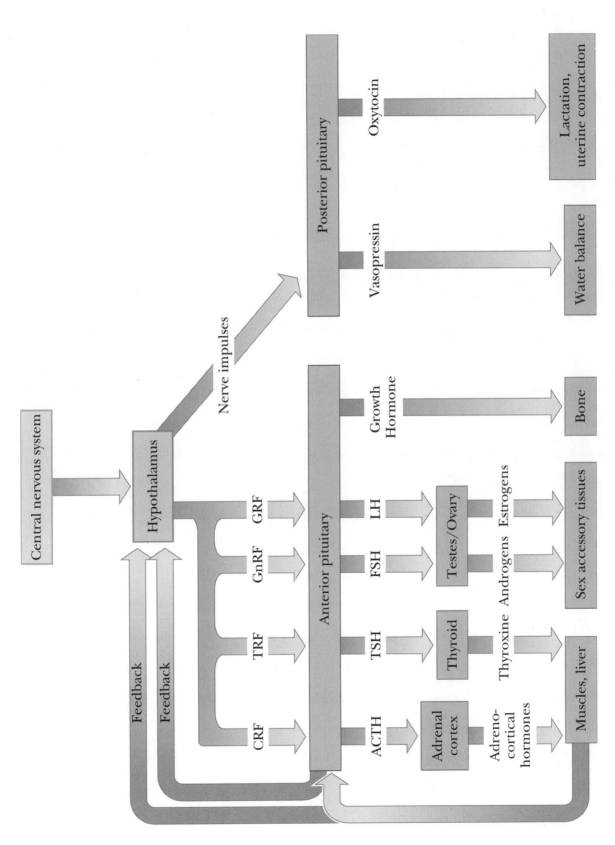

Figure 24.7 Hormonal control system

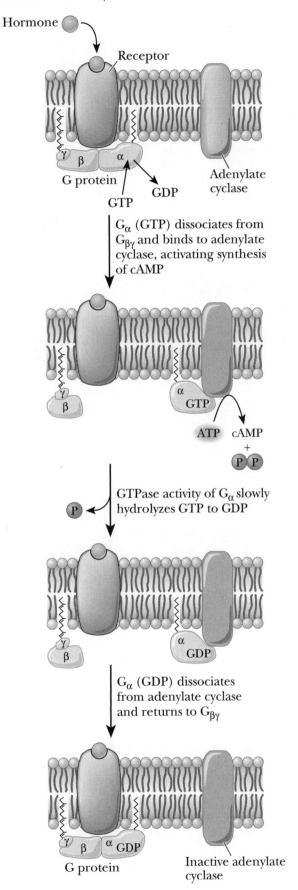

Figure 24.9 Activation of adenylate cyclase by hetero-trimeric G proteins

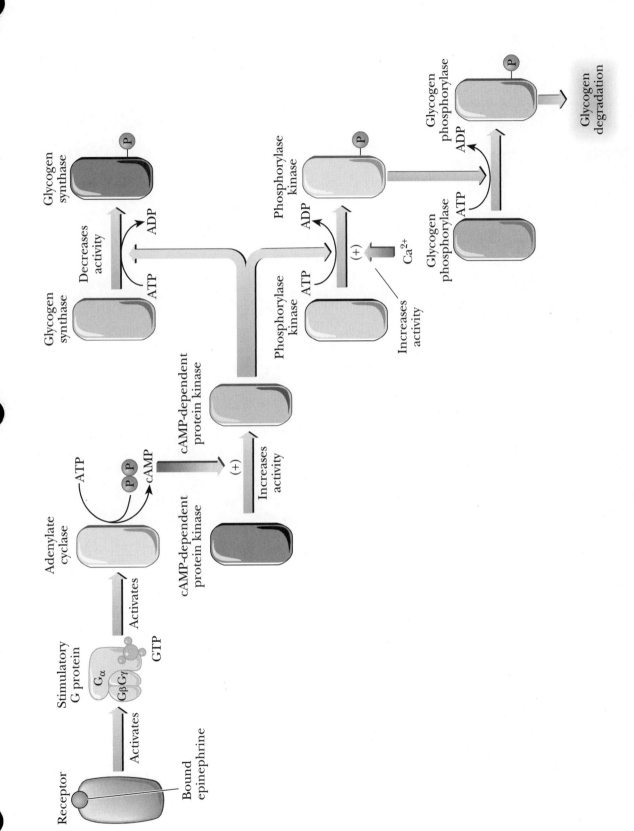

Figure 24.14 Epinephrine binding to receptor

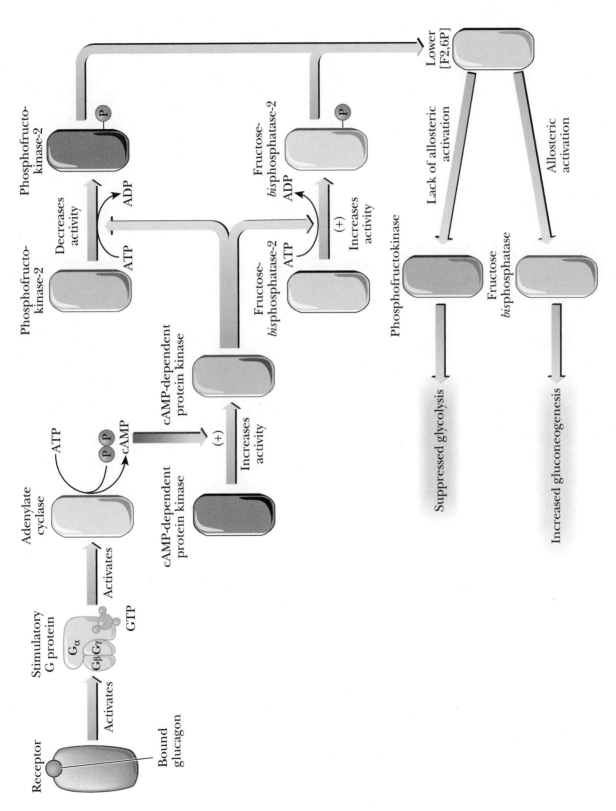

Figure 24.15 Glucagon binding to receptor